28 Fortschritte der chemischen Forschung
Topics in Current Chemistry

π Complexes of Transition Metals

Springer-Verlag
Berlin Heidelberg New York 1972

ISBN 3-540-05728-5 Springer-Verlag Berlin Heidelberg New York
ISBN 0-387-05728-5 Springer-Verlag New York Heidelberg Berlin

 Library of Congress Catalog Card Number 51-5497. Printed in Germany. Satz, Offsetdruck und Bindearbeiten: Verlag A. Hain KG, Meisenheim/Glan

Contents

Theoretical Considerations for Cyclic $(pd)_\pi$ Systems

Priv.-Doz. Dr. Günter Häfelinger
Lehrstuhl für Organische Chemie der Universität Tübingen

Contents

I. Introduction

The term "aromaticity" or "aromatic character" is used in organic chemistry to indicate an extremely high thermodynamic ground-state stability, as well as high kinetic stability, which may be observed in some cyclic conjugated unsaturated compounds in comparison with appropriate non-cyclic unsaturated reference compounds [1-4]. This definition is vague because "aromaticity" denotes an *experimental* excess property whose magnitude depends on a more or less arbitrary selection of the reference compound.

A necessary *theoretical* condition for the occurrence of striking excess ground-state stability is the possibiliy of cyclic delocalization of π electrons in a molecular closed-shell ground-state configuration constructed from a set of cyclically overlapping p_π atomic orbitals. Hückel's famous rule [5] states that only those planar monocyclic conjugated unsaturated hydrocarbons containing $(4m + 2)$ π electrons fulfill this theoretical condition.

If one or more carbon atoms in the cycle are replaced by heteroatoms also contributing p_π orbitals to the cyclic π system, this effect may be treated by means of perturbation theory. The electronegativity difference between the heteroatom and the neighbouring carbon atoms ensures that the cyclic delocalization of π electrons will be disturbed, as an electronegative heteroatom acts as an electron trap and an electropositive heteroatom as a repulsive barrier. But at the same time, for even-membered neutral or odd-membered negatively charged cycles, the total π-electron energy is increased because of the increased π-electron attraction by a more electronegative heteroatom. As a result of both effects, the increased π-electron energy of heterocyclic π systems, as, for example, in pyridine, is not caused by increased cyclic π-electron delocalization but by the stabilizing inductive effect of the heteroatom. In addition, Hückel's rule can no longer be applied for the prediction of closed-shell or open-shell ground-state configuration. Because of the reduced symmetry of the cyclic π system by the introduction of heteroatoms, the molecular orbitals no longer occur in degenerate pairs. Therefore any even number of π electrons will lead to a closed-shell ground-state configuration (see p. 29).

Conjugated unsaturated heterocyclic compounds may be termed heteroaromatic if their corresponding π-electron model can be derived from an aromatic cyclopolyene by means of perturbation theory without too much disturbance of cyclic π-electron delocalization.

But what happens to the ground state properties of a cyclic π system if one replaces one or more p_π orbitals by d orbitals capable of forming π bonds? Can one still expect an unusual ground-state stabilization, i.e. aromaticity?

This question may be tackled by both a theoretical and an experimental approach. In this review both approaches are applied. First, the properties of π systems containing d orbitals are treated in a very general and idealized way by the use of the HMO π-electron model theory, the principles of which are assumed to be familiar to the reader [6,7]. Then, as a specific example, the properties of transition metal chelates containing aliphatic 1,2-diimine ligands are examined by means of the HMO theory and by group theoretical considerations for all the valence electrons.

II. Examples of Cyclic $(pd)_\pi$ Systems

Chemical bonds involving *d* orbitals can be formed only in molecules containing atoms of the third or higher rows of the periodic system. Depending on whether the *d* orbital involved in bonding originates from the valence electron shell (outer *d* orbital) or its preceding inner shell (inner *d* orbital) a distinction is made between two different types of bonds involving *d* orbitals which differ in their bonding strengths [8-10]. From Hartree-Fock calculations for atoms [11], it appears that outer *d* orbitals are too large to form strong bonds to neighbouring atoms whereas inner *d* orbitals are too small. But the size of *d* orbitals in molecules is very strongly affected by the kind of atoms attached [10] (see p. 30). Therefore atomic *d* orbitals are not a good representation of molecular *d* orbitals, so that, for any specific molecule, the extent of involvement of *d* orbitals in either σ or π bonds has to be considered separately both theoretically and experimentally.

1. Cyclic $(pd)_\pi$ Systems Using Outer *d* Orbitals

Some known monocyclic conjugated hetero π systems containing heteroatoms like phosphorous, sulphur or selenium, which may take part in the π system by means of either p_π or outer d_π orbitals, are shown in Fig. 1.

Among the typically organic heterocycles of Fig. 1, the 1-methylphosphole [12] *1* does not necessarily contribute d_π orbitals to the π-system besides its p_π lone pair [13]. The same holds for the phosphabenzene derivatives *2a* [14] and *2b* [15]. But in the phosphorine derivatives *3a* [16], *3b* [17], *3c* [18] and *3d* [19], as well as in 1,1,3,3-tetraphenyl-1,3-diphosphabenzene [20] *4* and 1-phenyl-1H-phosphepine-1-oxide [21] *5* which contain tetrahedral phosphorous atoms, cyclic conjugation can only be achieved by the use of outer *d* orbitals. The resultant bonding situation has been treated theoretically by Mason [22] using HMO theory.

The extent of sulphur's *d* orbital contribution to the π system in addition to that of one p_π lone electron pair was greatly overemphasized in the early HMO treatment [23] of thiophene *6*. But more advanced calculations show that

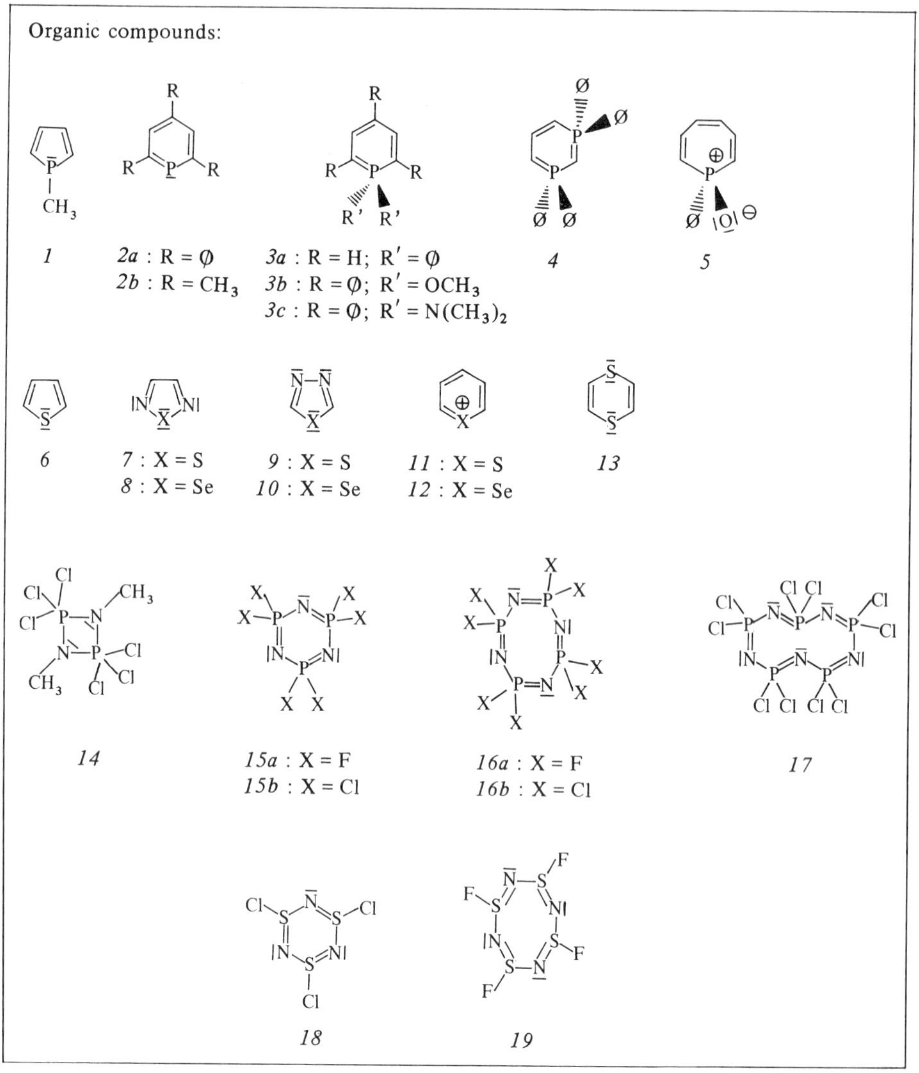

Fig. 1. Monocyclic conjugated hetero π systems in which outer d orbitals may be involved in cyclic conjugation

even a small d_π orbital contribution has quite a large effect on the calculated electronic properties [24-27].

The same conclusion holds true in principle for the thiadiazoles *7* [28] and *9* [29] or the selenadiazoles *8* [30] and *10* [31], whose structures are very accurately known experimentally, but which have been studied theoretically only by means of HMO [32,33] or PPP π-electron theories [34] without any specific examination of *d* orbital participation. Even less is known about the participation of *d* orbitals in conjugation in the 1,4-dithiadiene [24] *13* which is non-planar [35]. In the positively charged thiapyrylium ion [36,37] *11* or selenapyrylium ion [38] *12* the contribution of *d* orbitals to the π system may be more important [27,39], as in the other sulphur compounds listed here.

Of these organic unsaturated heterocyclic compounds with *d* orbital centers, mainly five-, six- and seven-membered monocyclic ring systems are known. But the typical inorganic ring systems listed in Fig. 1 contain planar four-membered *14* [40], six-membered *15a* [41], *15b* [42] and *18* [43], eight-membered *16a* [44] and nearly planar ten-membered *17* [45] rings, as well as non-planar eight-membered *16b* [46,47] and *19* [48], twelve-membered $[NP(NCH_3)_3]_6$ [49] and sixteen-membered ring systems $[NP(OCH_3)_2]_8$ [50].

The extent of *d*-orbital contribution to the π system, apart from that of a p_π orbital in the thiazyl halides [51,52] *18* and *19*, is not precisely known. But it is surely less [9] than in the phosphonitrilic halides *15*, *16* and *17*, whose bonding properties have been very extensively studied by Craig *et al.* [53-58]. There is no doubt that d_π orbitals of the tetrahedral phosphorous atoms are involved strongly in π-bonding, but Craig's formulation of a cyclic delocalized $(pd)_\pi$ system has been questioned by Dewar *et al.* [59], who suggested only three-centered delocalized bonds (*"island structures"*) without extensive cyclic delocalization. The use of outer *d* orbitals in bonding, especially in inorganic compounds of the type given in Fig. 1, was recently reviewed by Mitchell [9].

2. Cyclic $(pd)_\pi$ Systems Using Inner *d* Orbitals

The participation of inner *d* orbitals in bonding may be observed with transition metal atoms of the third or higher rows of the periodic table. If these transition metal atoms or ions are coordinated by heteroatoms of ligands which are themselves connected by conjugated multiple bonds, one obtains chelate rings of different ring sizes which may form cyclic $(pd)_\pi$ systems. Some examples of such compounds with different coordinating heteroatoms are collected in Fig. 2.

The *four-membered ring chelates* have not yet been studied with respect to *d*-orbital participation in cyclic conjugation. The amidines seem to form polymeric complexes [60] instead of monomeric chelate rings *20*; four-membered chelate rings have been formulated for diazoaminobenzene nickel (II) chelates [61], but without the proof of direct determination of structure. The sterically disadvantageous four-membered ring *21* is also avoided in transition metal salts of carbonic acids in favour of dimeric [62] or polymeric structures [63]. Structure determinations by X-ray prove that N, N-dialkyldithiocarbamate ligands [64] form planar chelate rings *22* with different transition metal ions ($R = N(C_2H_5)_2$; $M = (Ni^{2+})$ [65]; (Cu^{2+}) [66]; (Zn^{2+}) [67]).

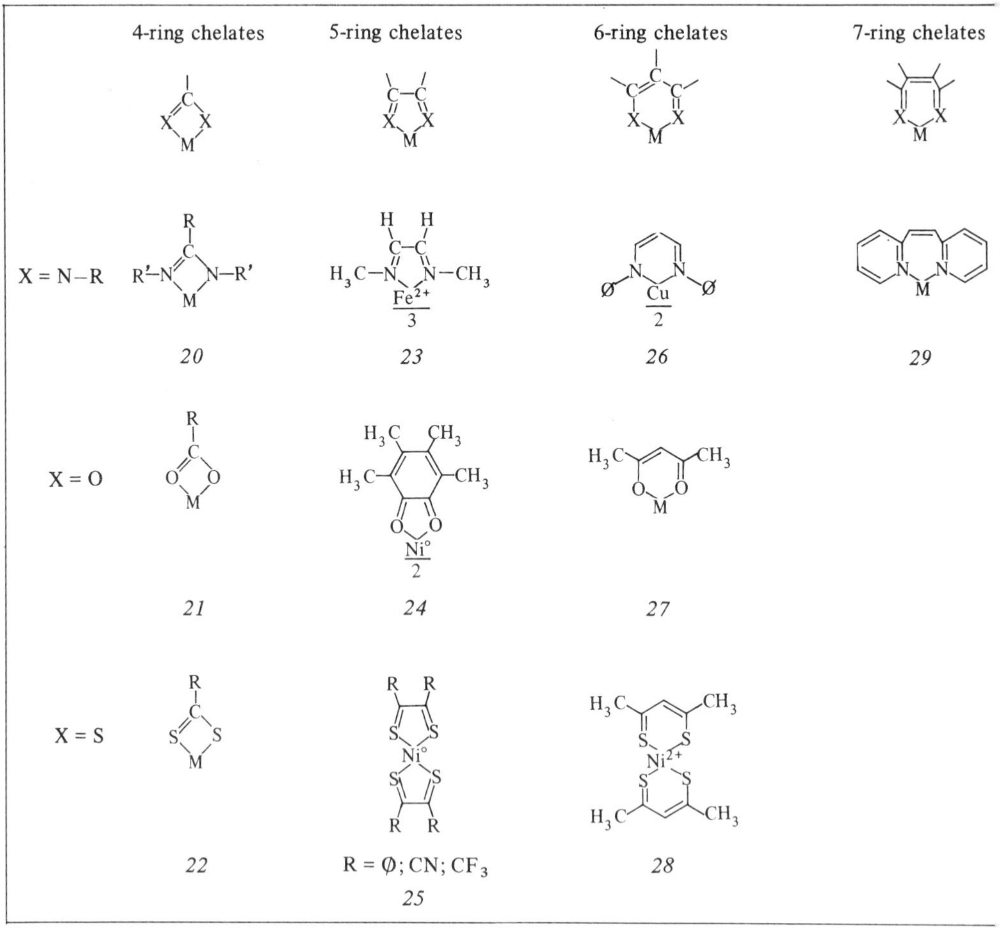

Fig. 2. Examples of transition metal chelates with unsaturated-ligand π systems which may use inner *d* orbitals for cyclic $(pd)_\pi$ bonding

The five- and six-membered transition metal chelate rings contain examples of compounds which have been most extensively studied, theoretically as well as experimentally. The tris-(glyoxalbis-N-methylimine)iron (II) chelate *23* synthesized by Krumholz [68] is the structurally simplest example of a vast number of intensely coloured chelates containing different 1,2-diimine ligand systems which have been recently reviewed [69]. $(pd)_\pi$-Bonding was thought to be so important for this type of chelates that it could explain their properties by an aromatic [70] or quasiaromatic [71,72] behaviour. These types of chelates will be considered specifically in Chapt. V.

Bis-duroquinone-nickel(0) [73] *24* is one of the few examples of five-membered ring chelates coordinated by oxygen. Coordination by sulphur leads to several transition metal chelates of type *25*, obtained from dithioketones or ethylene-1,2-dithiolates, which have been classified by Schrauzer [74] as "coordination compounds with delocalized ground states" [75,76] because they show extensive π-electron delocalization, including metal d_π orbitals [77].

The *bis-malodianil-copper(II)* [78] *26* is one example of six-membered ring chelates coordinated by nitrogen for which Daltrozzo [79] showed by light-absorption spectroscopy and NMR spectroscopy that no cyclic delocalization is observable across the transition metal.

The *acetylacetonate chelates* [80] *27* form an extensively studied class of complexes for which the suggestion that cyclic conjugation should lead to aromatic stability was applied historically for the first time [81,82]. However, Musso *et al.* [83,84] showed by analysis of vibration spectra that the π bonds in the chelated ligands are completely delocalized and the use of a mesityl substituent in position 3 as an indicator for diamagnetic ring currents showed no diamagnetic anisotropy comparable to that in benzene. They therefore discarded the concept of cyclic delocalization and aromatic character in these compounds.

Dithioacetylacetonate-chelates with Ni^{2+} and Co^{2+} of the form *28* have only lately been synthesized [85] and have not been treated theoretically.

Very little has been known till now about the properties of seven-membered ring chelates. Chelates with the cis-1,2-dipyridylethylene ligand *29* are being studied by the group of E. Bayer [86].

In addition to the chelates shown in Fig. 2, there might be any combination of the different coordinating ligand atoms, i.e. N,O,S connected in one ligand molecule, as well as different kinds of exocyclic or connecting groups, which would involve a tremendous number of possible structures. Whereas, for theoretical considerations, it is desirable to keep the molecule as small as possible and to study only the main important features, for chemical applications, i.e. in analytical chemistry, one tries to vary the molecular structure so as to influence certain desired properties, for example, solubility or specificity for certain ions.

Chelate rings coordinating through sulphur atoms may have contributions from sulphur outer d orbitals as well as from the central transition metal inner d orbitals. Therefore, it is also necessary to study the properties of cyclic π systems which contain more than one d-orbital center.

III. Theoretical Considerations for Idealized $(pd)_\pi$ Systems

1. Localized and Delocalized Molecular Orbital Description of Covalent Bonds

The formation of covalent chemical bonds in a molecule is quantum mechanically described by the overlap of atomic orbitals (AO) centered at different atoms which leads to one-electron molecular orbitals (MO) [87,88]; these may be classified as bonding, nonbonding or antibonding MO's depending on whether their energy is less, equal to or higher than that of the constituent AO's. Each of these MO's is completely delocalized about all the constituent atoms at the different atomic centers and may only be occupied by two electrons of opposite spin. The ground-state wave function is described by a Slater determinant of these delocalized MO's. Thus the concept of delocalization seems to be built in necessarily by the use of MO-treatment.

Making use of the known symmetry properties of the molecule, however, it is possible to transform the Slater determinant of delocalized MO's by a unitary transformation into another Slater determinant containing a set of equivalent MO's corresponding to localized bonds [89,90]. While this unitary transformation does not affect the value of the total electronic ground-state energy, the orbital energy values are shifted. There are an infinite number of possible unitary transformations of the set of delocalized MO's, i.e. the transformation to equivalent MO's is not unique but permits an interpretation of a given phsical situation in terms of bonds, a symbol which is very familar to chemical thinking. This transformation into localized bonds is possible for σ electrons of a planar molecule, but not for the π electrons, which remain delocalized about the framework of p orbitals with π symmetry [91]. Therefore the concept of delocalization is associated to a much higher degree with π-electron systems than with σ electrons [92]. The chemically observable effect associated with delocalization is conjugation, i.e. the ease transmittance of electronic effects [93] as well as strong molecular magnetic anisotropy.

Determinantal MO's may be obtained by a large number of computational methods based on Roothaan's self-consistent field formalism [94] for solving the Hartree-Fock equation for molecules which differ in degree of sophistication as regards the completeness and kind of the set of starting atomic wave functions, as well as the completeness of the Hamiltonian used [95]. So a chain of various kinds of approximations is available for calculations: starting from different ways of non-empirical *"ab initio"* calculations [96], via semiempirical methods for all-valence electrons with inclusion of electronic interaction [95,97,98] (CNDO [99], INDO [100], PNDO [101]) or with neglect of electronic interaction ("extended Hückel method") [102] and ending with π-electron models either with inclusion of electronic interaction [103,104] (PPP) [105,106] or with neglect of electronic interaction [6,7] (HMO) [5]. The adoption of the HMO theory for the consideration of π electrons in this article implies that these "theoretical" considerations refer, not to real molecules, but to a model selected in a consistent manner which retains the name of a real molecule only for convenience. It has the advantage of allowing a comparison between predictions for a number of structurally related molecular models. If one then carefully checks whether any correlation exists between experimental values and the HMO-model prediction for the same property, one may be able to explain and interpret trends in the behaviour of classes of compounds.

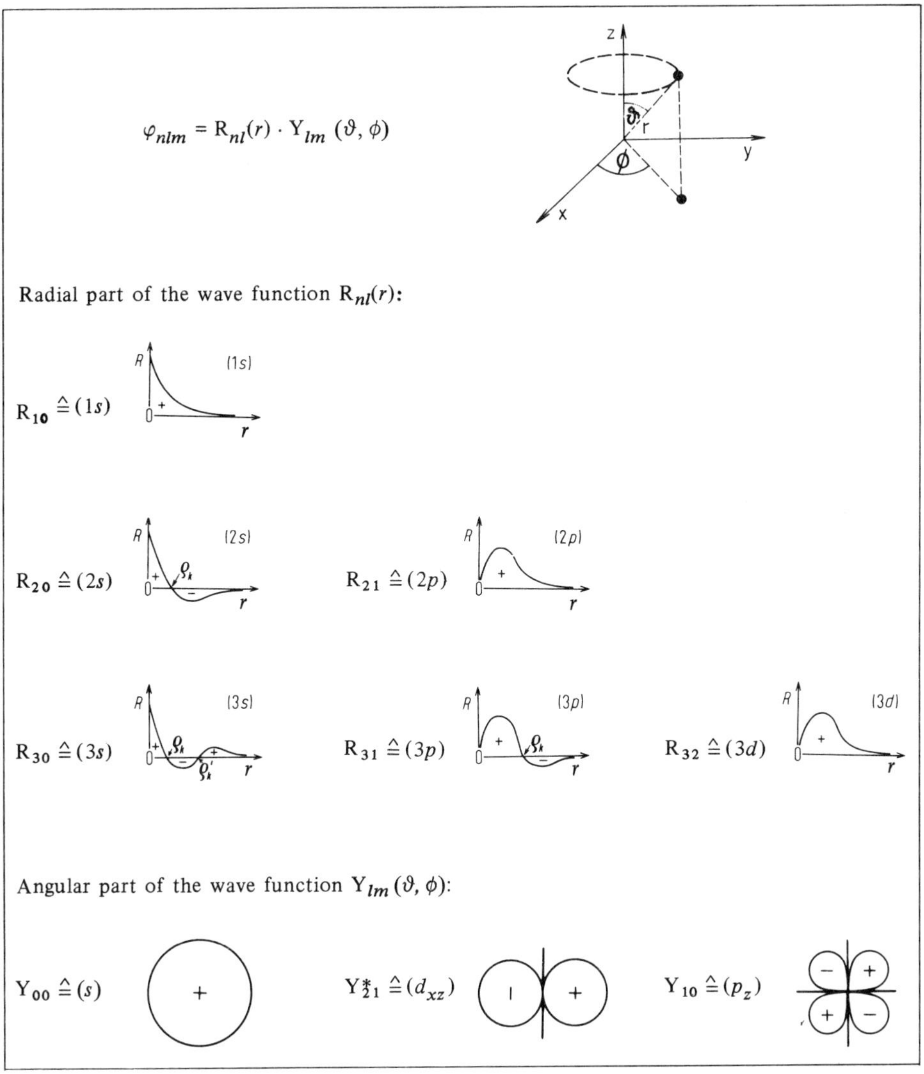

Fig. 3. Qualitative graphical representation of atomic wave functions (r, ϑ, ϕ = spherical coordinates, n, l, m = quantum numbers, ρ_K = nodes in radial functions, *) = real linear combination for $m = \pm 1$) (By permission the 9 diagrams of the lower part of this figure have been reproduced from E. Heilbronner and H. Bock: Das HMO-Modell und seine Anwendung. Grundlagen und Handhabung, p. 36/37. Weinheim: Verlag Chemie GmbH 1968)

Fig. 3 shows a qualitative graphical representation of hydrogen-like wave functions for one-electron atoms which have to be replaced for many-electron atoms at least by Slater-type [107] analytical wave functions φ_{nlm} (1) which are approximate as they contain no nodes in the radial part R_{nl}.

$$\varphi_{nlm} = R_{nl}(r) \cdot Y_{lm}(\vartheta, \phi)$$
$$R_{nl}(r) = N_{nl} \cdot r^{n-l} \cdot e^{-\frac{Z-\sigma}{n} \cdot \frac{r}{a_H}} \tag{1}$$

Meaning of the symbols: n, l, m = quantum numbers
r, ϑ, ϕ = spherical coordinates
N_{nl} = normalization constant
Z = charge of nucleus
σ = atomic screening constant
a_H = 0.529 Å = Bohr's atomic radius.

This radial part is the characteristic unknown for any atom of the periodic system and may be obtained by a variety of computational approaches. It may be determined, for example, numerically by Hartree-Fock calculations or by direct minimalization of Slater orbital exponents [108] ($\zeta = \frac{Z-\sigma}{n}$).

The angular part of the wave function $Y_{lm}(\vartheta, \phi)$ is independent of the main quantum number n and the same is true of any kind of atomic wave function at any level of sophistication. Therefore the spatial distribution and symmetry properties of any atomic s, p, or d orbital is the same for each atom of the periodic table or for different shells of the same atom.

For the formation of a chemical bond between atoms, two main conditions have to be fulfilled. The principle of maximum overlap [109] states that strongest bonds will be formed when the atomic orbital energies do not differ too much and the atomic orbitals have the correct symmetry to give nonvanishing values for the overlap integral S_{AB} (2).

$$S_{AB} = \int_{\tau} \varphi_A^* \varphi_B \, d\tau \tag{2}$$

The second condition may be fulfilled by a consideration of the angular part of the wave functions, which leads to a symmetry classification of bond types between atoms. Fig. 4 shows all types of bonding possible by combinations of s, p and d orbitals centered at two different atomic centers A und B. The types of bonds formed may be classified by means of their symmetry behaviour as σ bonds whose wave functions are symmetrical with respect to rotation around the connecting bond axis, as π bonds whose wave functions are antisymmetrical with respect to a nodal plane containing the bond axis, and as δ bonds having two perpendicular nodal planes with the bond axis lying in the intersecting line, although the last are of very little importance.

Table 1 gives numerical values of overlap intergrals (2) for different $(pp)_\pi$

Table 1. *Values of $(pp)_\pi$ and $(pd)_\pi$ overlap integrals (2) for Slater orbitals* [107] *and Clementi orbitals* [108]

	Slater orbitals						Clementi orbitals			
π bond (A – B)	$\rho^{a)} \cdot \frac{a_H}{r_{AB}}$	$\tau^{a)}$	r_{AB}^{min} [Å]	$S_\pi(r_{min})$	r_{AB}^{max} [Å]	$S_\pi(r_{max})$	$\rho^{a)} \cdot \frac{a_H}{r_{AB}}$	$\tau^{a)}$	$S_\pi(r_{min})$	$S_\pi(r_{max})$
C(2*p*)-C(2*p*)	1.625	0.00	1.29	0.29	1.58	0.18	1.5679	0.00	0.32	0.19
N(2*p*)-C(2*p*)	1.788	0.09	1.24	0.26	1.48	0.16	1.7425	0.10	0.27	0.17
O(2*p*)-C(2*p*)	1.950	0.17	1.18	0.22	1.43	0.14	1.8973	0.17	0.25	0.15
C(2*p*)-P(3*p*)	1.613	0.01	1.60	0.24	1.84	0.17	1.5984	-0.02	0.25	0.18
C(2*p*)-S(3*p*)	1.729	-0.06	1.57	0.23	1.80	0.15	1.6976	-0.08	0.24	0.16
N(2*p*)-Fe^0(3*p*)	3.433	-0.43	1.80	0.02	2.00	<0.01	3.0882	-0.38	0.02	0.01

Table 2. *Values of $(pd)_\pi$ overlap integrals (2) for Slater orbitals* [107] *and Clementi orbitals* [108]

	Slater orbitals						Clementi orbitals			
π bond (A – B)	$\rho^{a)} \cdot \frac{a_H}{r_{AB}}$	$\tau^{a)}$	r_{AB}^{min} [Å]	$S_\pi(r_{min})$	r_{AB}^{max} [Å]	$S_\pi(r_{max})$	$\rho^{a)} \cdot \frac{a_H}{r_{AB}}$	$\tau^{a)}$	$S_\pi(r_{min})$	$S_\pi(r_{max})$
Outer *d* orbitals:										
C(2*p*)-P(3*d*)	1.354	0.20	1.60	0.41	1.84	0.33	1.0191	0.54	0.27	0.27
C(2*p*)-S(3*d*)	1.463	0.11	1.57	0.41	1.80	0.34	1.1857	0.32	0.45	0.40
N(2*p*)-P(3*d*)	1.517	0.29	1.45	0.40	1.65	0.35	1.1936	0.61	0.26	0.25
N(2*p*)-S(3*d*)	1.625	0.20	1.50	0.38	1.70	0.30	1.3603	0.42	0.27	0.25
Inner *d* orbitals										
N(2*p*)-Fe^{3+}(3*d*)	2.075	-0.09	1.80	0.13	2.00	0.07				
N(2*p*)-Fe^{2+}(3*d*)	2.017	-0.03	1.80	0.14	2.00	0.10				
N(2*p*)-Fe^0(3*d*)	2.017	-0.03	1.80	0.14	2.00	0.10	2.8218	-0.32	0.030	0.015
N^+(2*p*)-Fe^-(d^2sp^3)	1.987	0.07	1.80	0.16	2.00	0.11				
N(2*p*)-Ni^0(3*d*)	2.233	-0.13	1.80	0.09	2.00	0.05	3.0467	-0.38	0.021	0.010
N^+(2*p*)-Ni^-(dsp^2)	2.204	-0.04	1.80	0.10	2.00	0.07				

a) For Slater-type orbitals the orbital exponent ζ is given by $\zeta = \frac{Z-\sigma}{n}$. The use of overlap integral tables requires that two parameters ρ and τ be calculated for each bond between the two atoms A and B: $\rho = \frac{1}{2}(\zeta_A + \zeta_B)\frac{r_{AB}}{a_H}$; $\tau = \frac{\zeta_A - \zeta_B}{\zeta_A + \zeta_B}$.

bonds whith maximum and minimum values of bond distances taken from empirical HMO π-bond order-bond length relations [110] (Table 1 see p. 11)

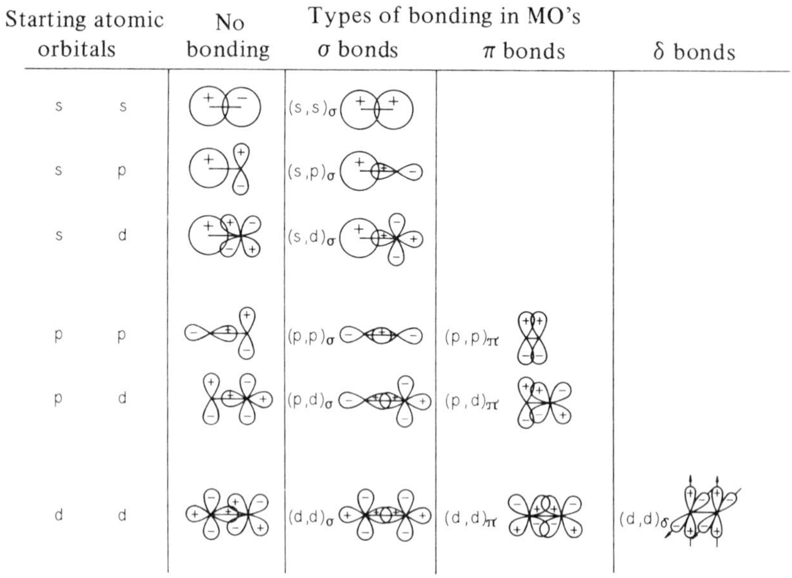

Fig. 4. Types of optimum chemical bonds possible by overlap of *s, p* and *d* orbitals centered at two atoms A and B

The values of the overlap integrals given for both Slater orbitals [107] and Clementi orbitals [108] have been taken from Mulliken's table [111]. They lie in the range from 0.32 to 0.14, showing good agreement for Clementi orbitals and those obtained by Slater's rules.

The overlap integrals for $(pd)_\pi$ bonds [112], given in Table 2, show some disagreement because Slater's rules do not apply too well for *d* orbitals. $(pd)_\pi$ overlap integrals using outer *d* orbitals seem to be at least as big as those of $(pp)_\pi$ systems or even larger, whereas $(pd)_\pi$ overlap intergrals for inner *d* orbitals, while smaller, are by no means negligible [8]. The use of inner *d* orbitals of transition metal ions produces an important uncertainty in the determination of screening constants which will be discussed later (p. 28) (Table 2 see p. 11).

Fig. 5 shows the usual graphical representation of the spatial distribution of atomic orbitals which is obtained as the square of the angular part of the wave functions provided with the sign of the angular part of the wave functions.

For the following discussion the cartesian coordinate axes of an atom containig *d* orbitals will always be directed so as to point as far as possible towards the atoms which are directly bound neighbours. Then, in the case of octahedral symmetry, only p_x, p_y and p_z orbitals, as well as d_{xy}, d_{xz} and d_{yz} orbitals, can be used for π-bonding of the type $(pp)_\pi$, $(pd)_\pi$, or $(dd)_\pi$.

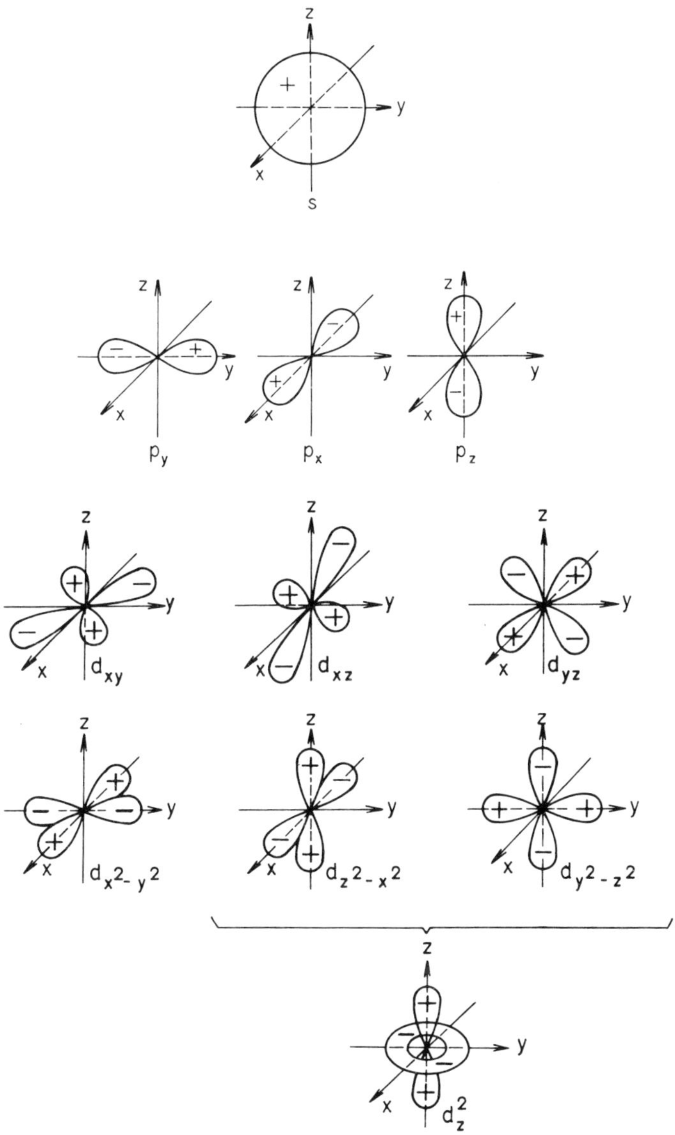

Fig. 5. Graphical representation of the spatial distribution of atomic *s*, *p* and *d* orbitals (square of real angular functions with indication of the sign of the angular functions)

In a planar molecule with the *xy* plane as the plane of the molecule, only p_z orbitals od d_{xz} and d_{yz} orbitals are able to form π bonds. All the other orbitals shown in Fig. 5 can only form σ bonds of the kind depicted in Fig. 4.

These σ bonds, which may be constructed either from delocalized σ AO's or from localized hybrid atomic orbitals, are neglected in a π-electron approximation for the delocalized π system.

2. HMO Criteria for Aromaticity in Cyclic $(pp)_\pi$ Systems

A comparison of various properties of the HMO model for linear and monocyclic conjugated $(pp)_\pi$ systems made it possible to derive numerical values for a set of upper or lower bounds differentiating theoretically between typical aromatic or olefinic behaviour in cyclic π systems [113]. The most important of these are collected in (3).

A cyclopolyene is "aromatic" if:		values for benzene:	
(a) $\lvert\Delta\epsilon\rvert$	$> 0.0\,\beta$	$\lvert\Delta\epsilon\rvert$	$= 2.0\,\beta$
(b) E^b_π/N	$> 1.273\,\beta$	E^b_π/N	$= 1.333\,\beta$
(c) DE^{cycl}	$> 0.727\,\beta$	DE^{cycl}	$= 1.012\,\beta$
(d) $L^{\min}_\mu$	$\geqslant 2.000\,\beta$	$L^{\min}_\mu$	$= 2.536\,\beta$

(3)

The notations and definitions used here are given in Table 3.

Condition (3a) is necessary to ensure a closed-shell ground-state configuration, whereas conditions (3b) and (3c) are minimum conditions for aromatic thermodynamic ground-state stability of π systems differing in their number and topological connection of p_π orbitals, as well as in the number of π electrons present. Condition (3d) is necessary to specify low aromatic reactivity.

But the striking excess property "aromaticity" might be even better characterized by values close to that of the standard aromatic compound, benzene, which are given in (3) for comparison.

The specific π-bond energy E^b_π/N, the cyclic delocalization energy DE^{cycl}, and the minimum localization energy $L^{\min}_\mu$, are functions of the number of p_π orbitals (N) as well as of the number of π electrons (n) which have maximum values only for certain numbers of π electrons and which differentiate very well between olefinic and aromatic behaviour. Therefore they are better suited for comparisons of different π systems than the properties most commonly used: the π-bond energy per π electron [5] E^b_π/n, which increases with increasing N for constant n and decreases with increasing n for constant N [113]; the delocalization energy [6] DE, which increases with N [113]; or the specific delocalization energy [114] DE/N, which does not differentiate clearly between olefinic and aromatic behaviour [113].

3. Peculiarities Due to Replacement of p_π Orbitals by d_π Orbitals

In this section no distinction is made between outer or inner d orbitals, and for simplicity it is assumed that the d-orbital center may be described by the

Table 3. *Notations and definitions of the HMO model theory* [6,7,113]

1) Notations:	
N	Number of atomic p_π orbitals
n	Number of π electrons
b_j	Occupation number of the π MO j
$\alpha_\mu = \alpha$	Empirical Coulombic integral parameter for a p_π atomic orbital at atom μ
$\beta_{\mu\nu} = \beta$	Empirical bond integral parameter between bonded atoms μ and ν
R_N	Ring constructed from N overlapping p_π orbitals
K_N	Linear chain of N p_π orbitals
2) Definitions:	
$\epsilon_j = \alpha + x_j\beta$	HMO orbital energy
$\lvert\Delta\epsilon\rvert$	Oribital energy difference between highest occupied MO and lowest unoccupied MO
$E_\pi^{tot} = \sum_{j=1}^{N} b_j\epsilon_j = n\alpha + \sum_{j=1}^{N} b_jx_j\beta$	Total π-electron energy
$E_\pi^b = E^{tot} - n\alpha = \sum_{j=1}^{N} b_jx_j\beta$	π-Bond energy
E_π^b/N	Specific π-bond energy
$DE = E_\pi^b - k \cdot 2\beta$	Delocalization energy; k = maximum number of double bond in one Kekulé structure
DE/N	Specific delocalization energy
$DE^{cycl} = (E^b(R_N) - E^b(K_N))_{n=const}$	Cyclic delocalization energy
$L_\mu = E^b(R_N) - E^b(K_{N-1})$	Localization energy
$p_{\mu\nu} = \sum_{j=1}^{N} b_jc_{j\mu}c_{j\nu}$	π-Bond order
$c_{j\mu}$	LCAO – MO coefficient
$F_\mu = \sqrt{3} - \sum_r p_{\mu\nu}$	Free valence
$q_\mu = \sum_{j=1}^{N} b_jc_j^2$	π-Electron density
$\pi_{\mu\mu} = \frac{\partial\rho_\mu}{\partial\alpha_\mu}$	Atom self-polarizability

same α value as the other p_π orbitals of the planar π system, and that the β value for a $(pd)_\pi$ bond may be the same as for a $(pp)_\pi$ bond of standard distance.

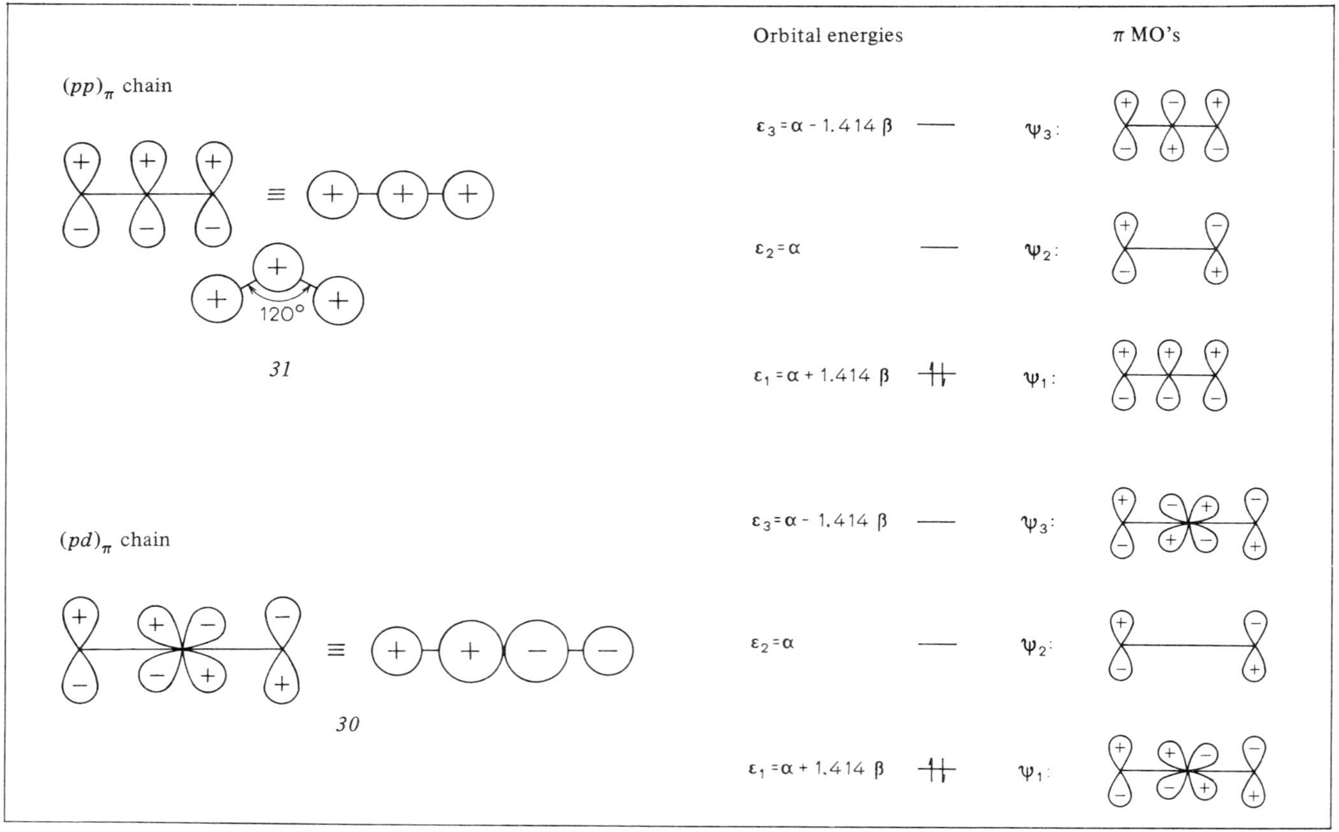

Fig. 6. HMO results for linear $(pp)_\pi$ and $(pd)_\pi$ polyene chains

3.1 Linear $(pd)_\pi$ Systems

If a d_π orbital (d_{xz} or d_{yz}) is directed between two p_z orbitals so as to give maximum π overlap *(30)*, the following differences may be noted with respect to a π system built only from *p* orbitals *(31)*:

a) As a d_π orbital has a preferred orientation in one plane, the $(pd)_\pi$ bond has to be linear, whereas in a $(pp)_\pi$ system the angle between three parallel *p* orbitals may have any arbitrary value (preferably 120°) which is allowed by the neglected σ skeleton without affecting the π overlap, as long as the molecule remains planar.

b) The d_π orbital has a second nodal plane perpendicular to the molecular plane. Therefore, for positive overlap, one *p* orbital has to reverse its sign, leading to a bonding π MO that has one additional node perpendicular to the molecular plane. But the π-electron distribution given by the square of the wave function is not affected by that change, nor is the conjugative property of this MO affected.

c) The second nodal plane of the *d* orbital allows an additional type of π-bonding *(32)* in which two *p* orbitals are arranged at right angles to each other as well as perpendicular to the molecular *xy* plane [115].

The result of an HMO calculation for the linear $(pd)_\pi$ system [7] *(30)* is compared in Fig. 6 with that for the allylic $(pp)_\pi$ system *(31)* and may be generalized as follows:

Within the limits of the above-listed approximations, a linear $(pd)_\pi$ system differs from a linear $(pp)_\pi$ system only in the symmetry behaviour of the MO's, which change sign at the *d*-orbital center. But the orbital energies, total π-electron energy as well as total π-electron occupation numbers, will be the same in both cases so that a closed-shell configuration may be expected for any even number of π electrons. The $(pd)_\pi$-bonding case *(32)* does not even differ in the symmetry behaviour of the π MO's. The only peculiarity of linear $(pd)_\pi$ systems is the preferred orientation in one plane caused by the *d* orbital with an angle of either 180° (case *30)* or 90° (case *32*) between the connecting σ bonds.

If the *d* orbitals are arranged in a way that allows no maximum overlap, the overlap integral (2) is reduced by the cosine of the angle of deviation γ in case *(30)* or by cos 2 γ for case *(32)*, as shown in Fig. 7. Therefore case *32* is more sensitive to angular deviations.

3.2 Cyclic $(pd)_\pi$ Systems

The replacement of a *p* orbital of a cyclic π system by one d_π orbital may result in two extreme directions of the *d* orbital, either radial or tangential orientation.

A radially directed *d* orbital shows the same symmetry behaviour as the replaced *p* orbital *33*. Therefore the properties of this cyclic $(pd)_\pi$ system are basically the same as in a π system containing only *p* orbitals, except for the

Case *30:* linear conjugation in molecular plane (xy)

Maximum overlap:

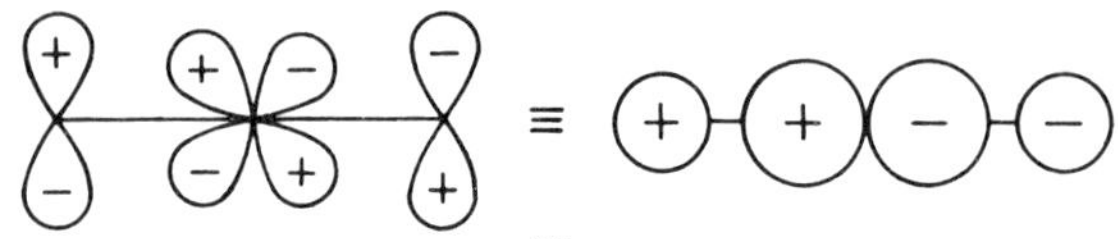

30

Reduced overlap: $S(pd)_\pi \sim \cos\alpha$

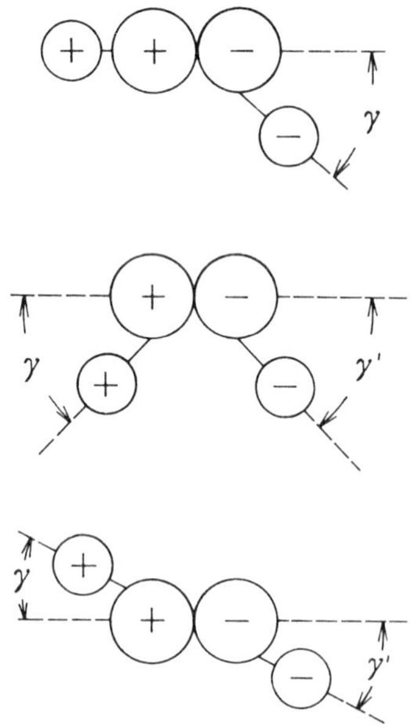

No overlap: $S\,(pd)_\pi = 0$

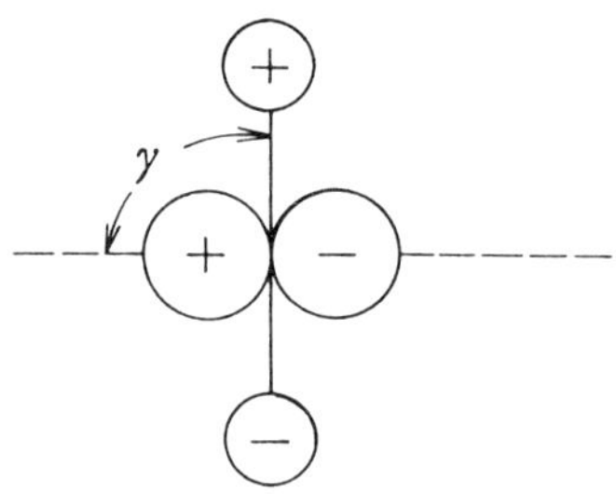

Fig. 7. Reduction of $(pd)_\pi$

Case *32:* conjugation perpendicular to molecular *xy* plane

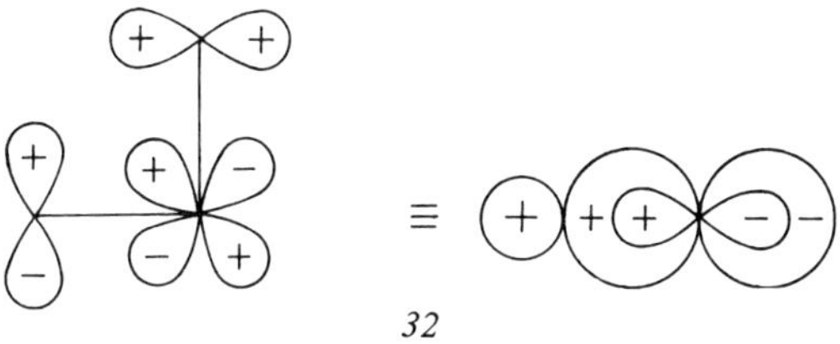

32

$S(pd)_\pi \sim \cos 2\gamma$

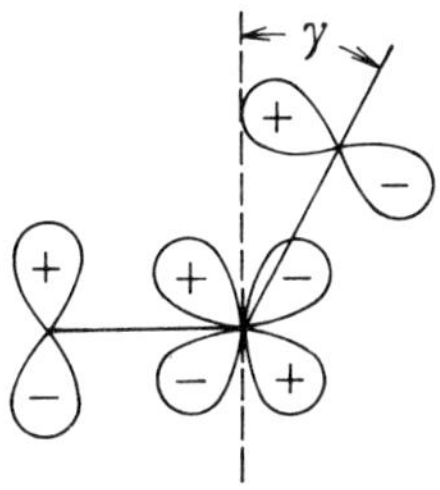

$S(pd)_\pi = 0$

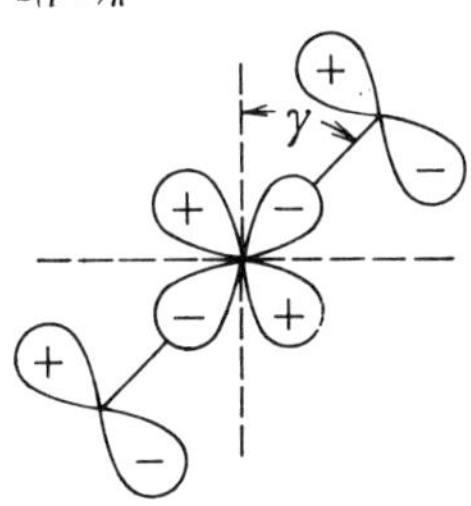

overlap by angular deviation

effect of reducing the overlap to the radial *d* orbital, which is dependent on the size of the cycle but which is neglected at the moment. The HMO orbital energies of this idealized Hückel-type $(pd)_\pi$ system may be given in closed form [5] (4) for different ring sizes.

$$\epsilon_j = \alpha + 2\beta \cdot \cos \frac{2j\pi}{N}; \quad j = 1, 2, \ldots, N \tag{4}$$

If the d_π orbital is directed tangentially, all MO's of the π system contain an additional node which is caused by the symmetry behaviour of the tangential *d* orbital. But, as this additional node is not located at a fixed position, one obtains a delocalized $(pd)_\pi$ system showing the same properties as a Möbius π system *34* which may be obtained by gradually twisting all *p* orbitals of a π system through a small angle in the same direction [116].

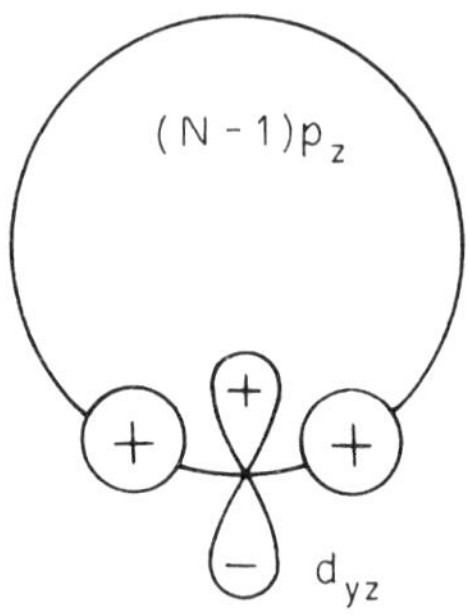

33

Hückel-type
$(pd)_\pi$ system
(radial *d* orbital)

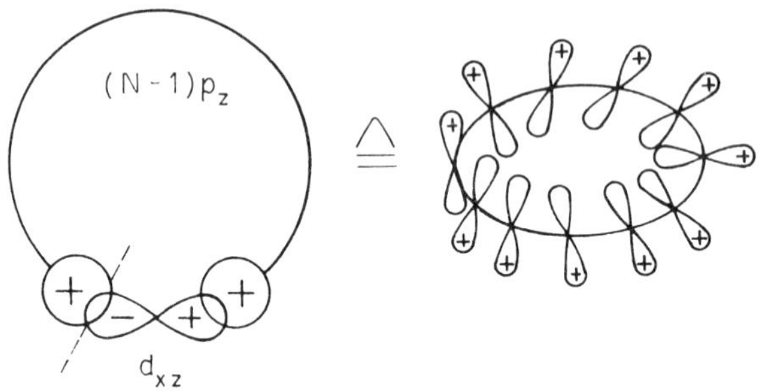

34

Möbius-type $(pd)_\pi$ system
(tangential *d* orbital)

The HMO-orbital energies for idealized Möbius-type $(pd)_\pi$ systems (9) may be given in closed form [116] (5).

$$\epsilon_j = \alpha + 2\beta \cdot \cos \frac{(2j+1)}{N}; \quad j = 0, 1, 2, \ldots, N-1 \tag{5}$$

But at a *d*-orbital center both types of $(pd)_\pi$-bonding are possible, generally with different probability because of differences in $(pd)_\pi$ overlap, which depends on the geometry of the cycle. If both radial and tangential *d* orbitals may be equally involved in π bonds, the situation is best decribed by an orthogonal linear combination of both *d* orbitals (6)

$$d^+ = \frac{1}{\sqrt{2}}(d_{yz} + d_{xz}); \quad d^- = \frac{1}{\sqrt{2}}(d_{yz} - d_{xz}) \tag{6}$$

which leads to a rotation of the d orbitals by 45° around the z axis *(35)*. The resulting bonding situation *(35)* no longer corresponds to a cyclic conjugated π system

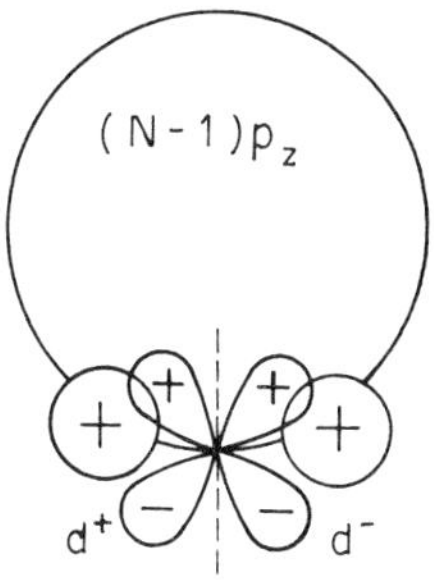

35

because the overlap between the two orthogonal d orbitals (6) is zero, but corresponds to $(N+1)$ linear polyene chains whose orbital energies are given[117] by (7).

$$\epsilon_j = \alpha + 2\beta \cos \frac{j\pi}{N+1}; \quad j = 1, 2, \ldots, N \tag{7}$$

The replacement of more than one p orbital of the cyclic π system by d orbital centers leads only in the case of tangential orientation of d_π orbitals to the following new situations: If each alternate p orbital in an even cycle is replaced by tangential d orbitals, the resulting situation *(36)* corresponds to Craig's [53–57] treatment of $(pd)_\pi$-bonding in the phosphonitrilic halides *15, 16, 17* whose idealized HMO orbital energy levels may be given in a closed form [54] (8) which

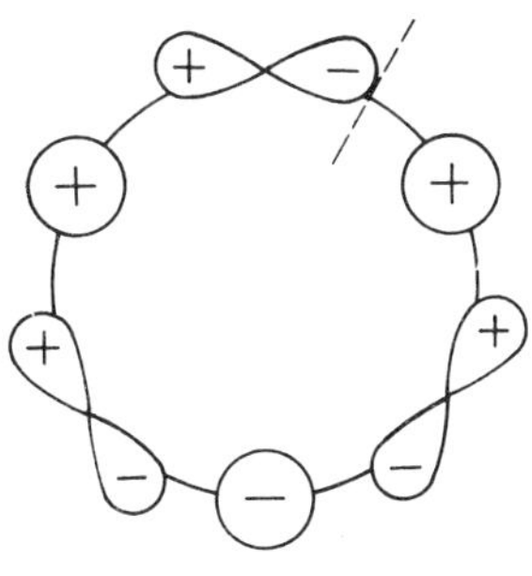

Craig-type $(pd)_\pi$ system

36

is also applicable for odd-membered rings.

$$\epsilon_j = \alpha + 2\beta \cdot \sin \frac{2j\pi}{N}; \quad j = 0, 1, \ldots, N-1 \tag{8}$$

We will characterize this case *(36)* as a Craig-type $(pd)_\pi$ system because Craig's original nomenclature [55] (homomorphic and heteromorphic $(pd)_\pi$ systems) is not applicable for all cases of $(pd)_\pi$-bonding considered here.

If only $(\frac{N}{2} - 1)$ or $(\frac{N}{2} + 1)$ p orbitals of an N-membered cycle are replaced by tangential d_π orbitals, one obtains idealized systems like, for example, *(37)*

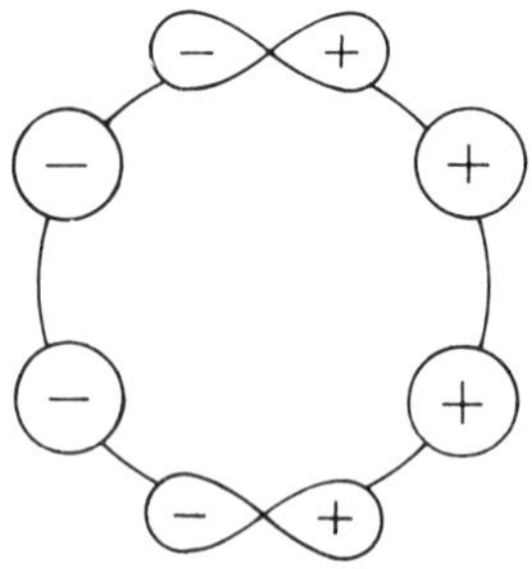

Mason-type $(pd)_\pi$ system

37

which corresponds to compounds *4* or *13* for whose HMO orbital energies the closed expression (9) was derived by Mason [22].

$$\epsilon_j = \alpha + 2\beta \cdot \sin \frac{(2j+1)\pi}{N}; \quad j = 0, 1, 2, \ldots, N-1 \tag{9}$$

If the actual molecular geometry in Craig-type *36* or Mason-type *37* planar $(pd)_\pi$ systems is such as to allow participation of both d_π orbitals, the bonding may again be described by means of the orthogonal linear combination *(6)* which leads to the interruption of cyclic conjugation at the d orbital centers *38*.

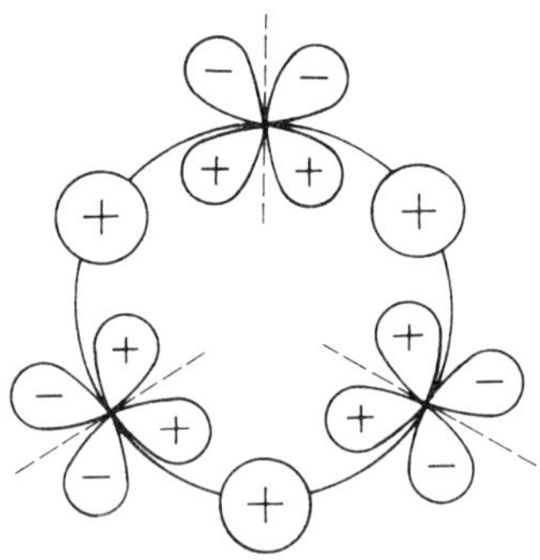

Dewar's island bond model

38

There may be formed independent sets of linear-polyene-like bonds, which have been termed by Dewar [59] island bonds [58], and whose HMO orbital energies may be obtained from (7).

4. HMO Properties of Cyclic $(pd)_\pi$ Systems

4.1 Orbital Energies and Optimum π Electron Occupation Numbers

The closed formula expressions (4), (5), (7), (8), and (9) allow very easy derivation of all HMO properties for rings of different size which depend on orbital energies. A comparative graphical representation of the HMO orbital energy levels for the different types of $(pd)_\pi$ systems is given for comparison in Fig. 8.

As in the case of Hückel-type π systems, it is possible to give a convenient mnemonic form for the closed formulas [118] which is based on the relation of cosine and sine functions to motion on a circle.

Hückel-type systems [118] (4)

Draw a circle of radius 2β around a midpoint corresponding to the α value and place inside it the regular polygon which corresponds to the desired cyclic π system so that one apex is always at the lowest point. The projection of the apices perpendicular to the connecting line from the midpoint of the circle to the lowest apex gives the numerical values of the orbital energies in units of β.

Möbius-type systems (5)

The polygon is placed inside the circle so that one edge is always in the lowest position.

Craig-type systems (8)

One apex of the polygon has always to have the same value as the center of the circle which corresponds to the α value.

Mason-type systems (9)

One edge of the polygon has always to be parallel to the line onto which the projection is made.

Fig. 8 shows that Hückel-type $(pd)_\pi$ systems (4) have one nondegenerate lowest bonding MO, and all the other bonding or nonbonding MO's occur in pairs of degenerate MO's. This leads by use of the aufbau principle only to closed-shell configurations for systems containing $(4m+2)\pi$ electrons, as suggested by Hückel's rule [5].

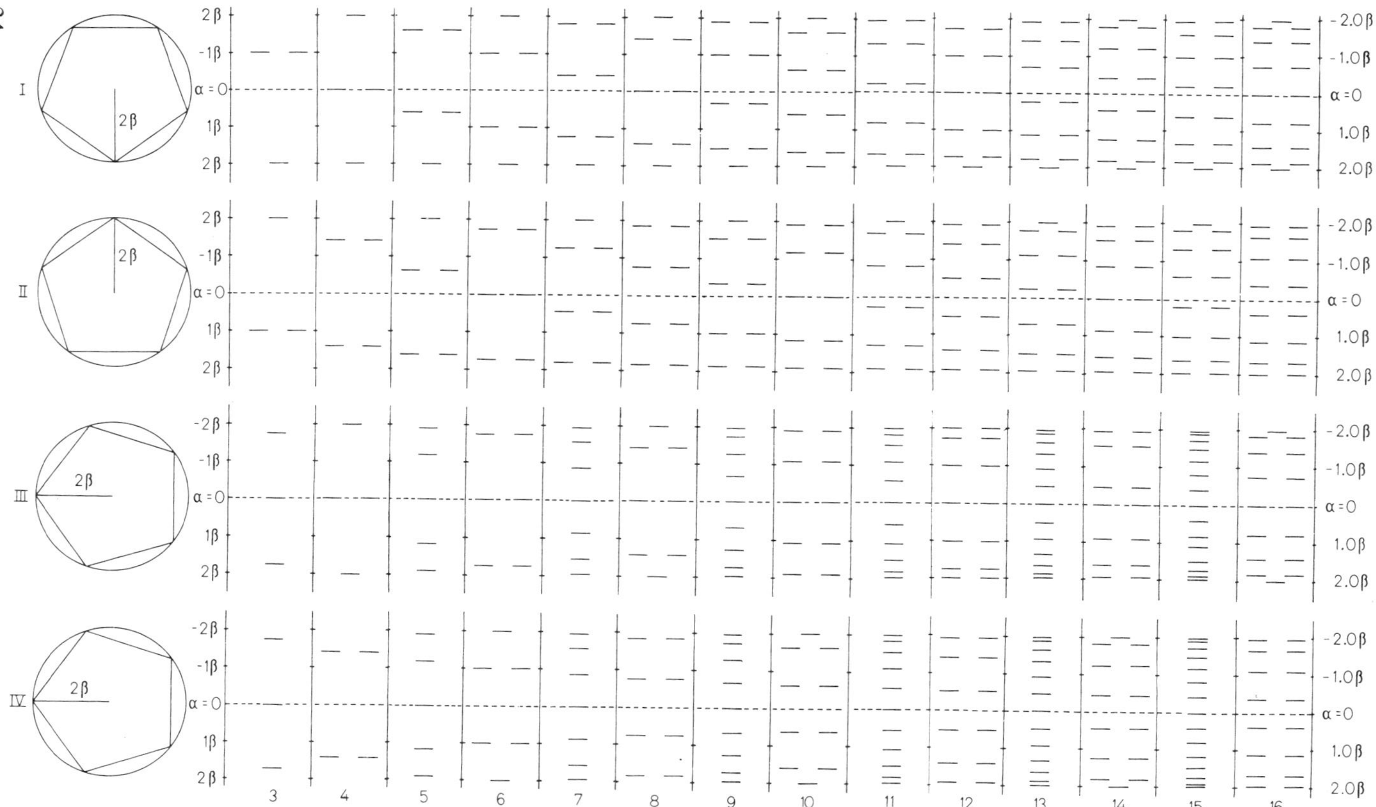

I = Hückel-type $(pd)_\pi$ systems (4) (radial d orbitals).

II = Möbius-type $(pd)_\pi$ systems (5) (one tangential d orbital).

III = Craig-type $(pd)_\pi$ systems (8) ($\frac{N}{2}$ tangential d orbitals).

IV = Mason-type $(pd)_\pi$ systems (9) [$(\frac{N}{2} \pm 1)$ tangential d orbitals].

Fig. 8. HMO orbital energies for idealized types of cyclic $(pd)_\pi$ systems with different ring size N

In Möbius-type $(pd)_\pi$ systems (5) all bonding or nonbonding MO's occur only in pairs of degenerate MO's. Therefore closed-shell configurations may only be obtained for numbers of $(4m)$ π electrons.

Even-membered Craig-type $(pd)_\pi$ systems (8) show alternatively the behaviour of a Hückel-type π system, if the arrangement of $\frac{N}{2}$ tangential d orbitals and $\frac{N}{2}$ p orbitals is possible without a node, as for example in *39*, or

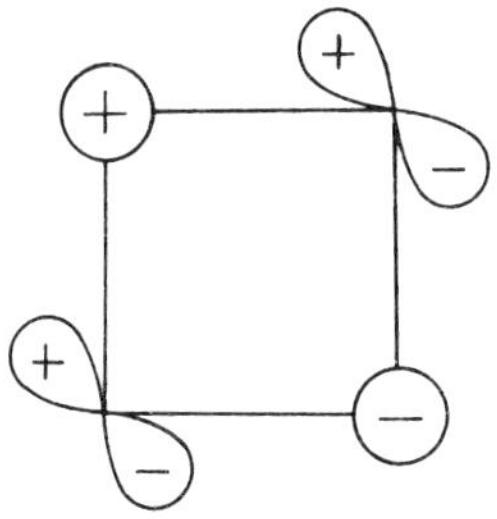

39

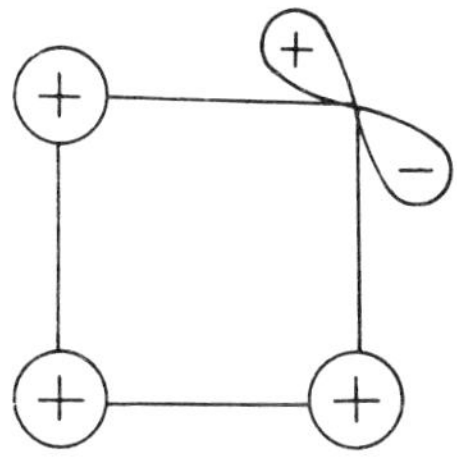

40

they show the pattern of a Möbius-type π system if it is not possible to arrange the orbitals without at least one nodal plane *36*. The same holds in principle for even-membered Mason-type π systems (9), as is seen from the representations *37* and *40* in comparison with Fig. 8.

For odd-membered Craig or Mason-type $(pd)_\pi$ systems, both formulas (8) and (9) lead to the same result: all MO's occur only singly with always one non-bonding MO in a pattern similar to that of odd-membered linear polyenes. Closed-shell configurations are therefore predicted for each even number of π electrons for both Craig and Mason-type $(pd)_\pi$ systems.

4.2 Ground State Stability Criteria

The graphical representation of specific π-bond energies (E^b_π/N) for the different types of $(pd)_\pi$ systems in Fig. 9 clearly shows the differences in stability.

Hückel-type π systems I show maximum values of specific π-bond energies for the three-membered ring occupied by two π electrons, for the five-, six- and seven-membered rings containing six π electrons, and so on, as predicted by Hückel's rule, the values of which decrease with increasing ring size. Minima are observed for the four-membered ring with four π electrons, for the eight-membered ring with eight π electrons, and so on. But these values are now increasing with increasing ring size, both kinds approaching the same limit of $1.273\ \beta$.

In Möbius-type π systems II, maximum values of E^b_π/N are observed for three-, four- and five-membered rings containing four π electrons and seven-,

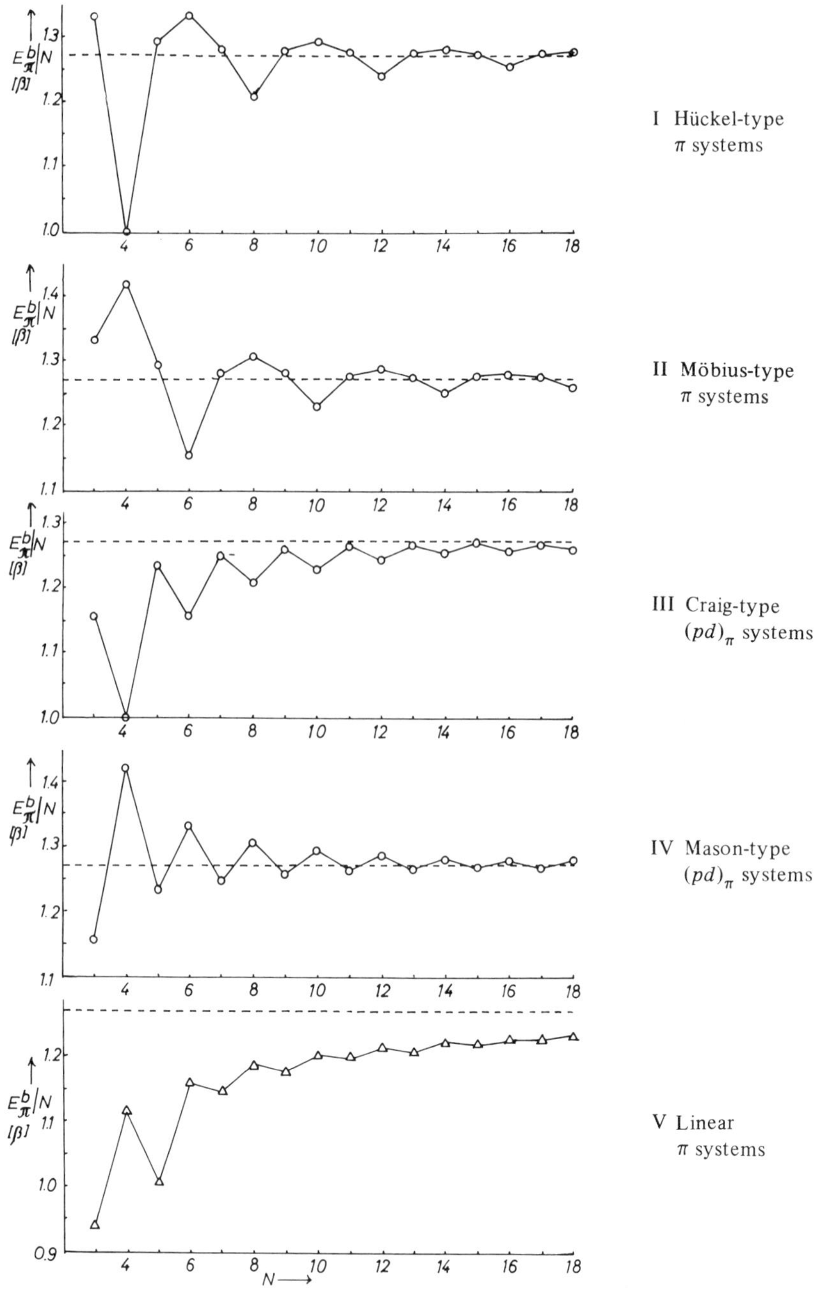

Fig. 9. Specific HMO π-bond energies E_π^b/N for idealized types of cyclic $(pd)_\pi$ systems in dependence on the ring size N for optimum π-electron occupation number

eight- and nine-membered rings containing eight π electrons and so on, again with decreasing values for increasing ring size. The minimum values are now obtained for $(4m+2)$-membered rings filled with the corresponding number of π electrons, i.e. the behaviour of Möbius-type π systems is the complete reverse of Hückel-type π systems.

For Craig-type $(pd)_\pi$ systems III, all values of the specific delocalization energy are below the borderline separating aromatic and olefinic behaviour.

The odd-membered Craig-type systems are more favourable than the even-membered, which correspond to the unfavourable cases of either Hückel or Möbius-type. Thus, in Craig-type $(pd)_\pi$ systems, no ring will show an extraordinary ground-state stability which may be termed aromatic. This prediction is the reverse of that of Craig [53–58] who based his conclusions on the delocalization energy per π electron. But there is agreement in the prediction that the stability will increase with increasing ring size, as is seen in the case of linear polyene systems V, and that the stability is greater than in a corresponding linear π system, except for four- and six-membered rings.

The Mason-type $(pd)_\pi$ systems IV represent energetically the most favourable situation. Only the values for the odd-membered rings, which are the same as in Craig systems, are below the aromaticity limit. All the even-membered rings exhibit the favourable values of either the Hückel or Möbius-type π systems.

The case of $(N+1)$ polyene chains by use of the linear combinations of the two d orbitals (6) may be read off from the graphical representation of the specific π-bond energy of linear polyenes in Fig. 9, but shifted by one unit to the left. The pattern of behaviour is similar to the case of Craig-type π systems but with slightly lower numerical values. So in this kind of bonding there is also no special aromatic stabilization to be expected.

Qualitatively, the same distribution of values as for the different $(pd)_\pi$ systems in Fig. 9 is observed for the cyclic delocalization energy, π-bond orders or reactivity indices, as localization energies or free valences [113], which does not yield additional information.

The results of the HMO π-electron considerations for idealized cyclic $(pd)_\pi$ systems may be summarized as follows:

For even-membered cyclic π systems containing d orbitals, only two basic types of π systems have to be considered. If it is possible to arrange the atomic orbitals with π symmetry graphically so that there is no change of sign between two neighbouring atoms, the system behaves like a Hückel-type π system with aromaticity for $(4m+2)\pi$ electrons. If the drawing shows a change of sign between neighbouring atoms, the system behaves like a Möbius-type π system with aromaticity for $(4m)\pi$ electrons.

For odd-membered $(pd)_\pi$ systems, an additional possibility exists which leads to a closed-shell configuration for any even number of π electrons, but without aromatic stability.

IV. Corrections of the Idealized HMO $(pd)_\pi$ Model

1. Deviations from Optimum $(pd)_\pi$ Overlap Caused by Geometry of Polygons

The replacement of a p orbital of a cyclopolyene by a d orbital, in either radial or tangential orientation, causes a deviation from the optimum overlap of the d_π orbital with the neighbouring p_π orbital which is proportional to the cosine of the angle of deviation between the most favourable linear arrangement and the actual angle of the regular polygon.

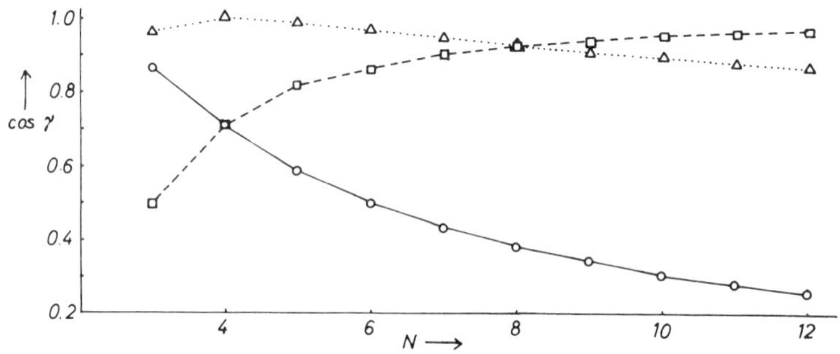

(——) for radial d-orbital orientation (– – –) for tangential d-orbital orientation
(------) for two perpendicular d orbitals (6)

Fig. 10. Plot of cosine of angle of deviation in dependence of size of regular polygons

Fig. 10 shows that, for the radial orientation of d orbitals, a strong reduction of $(pd)_\pi$-overlap intergrals with increasing size of regular polygons is to be expected, whereas the tangential orientation will become more favourable for larger rings. The linear combination (6) with a 90° difference in orientation of the two orbitals is least affected and decreases only slowly with increasing ring size. But for actual compounds, only regular polygon geometries up to $N \approx 9$ need be considered. The most important angles are those around 90° or 120°.

In an HMO calculation the reductions due to deviation from maximum $(pd)_\pi$ overlap may be introduced by the use of perturbation theory [119] instead of solving the secular determinant directly. As usual, the Hückel bond-intergral parameter $\beta_{\mu\nu}$ is assumed to be proportional to the overlap integral of the corresponding bond [6,7]. The reduction is therefore proportional in our case to the cosine of the angle of deviation γ (10).

$$\delta\beta_{\mu\nu} = (\cos\gamma - 1) \tag{10}$$

$$\delta E_\pi^b = 2p_{\mu\nu}\,\delta\beta_{\mu\nu} \tag{11}$$

The resultant change in π-bond energy by first-order perturbation theory is given in (11).

In Fig. 11 the correspondingly obtained corrected specific π-bond energies are plotted as a function of polygon size N.

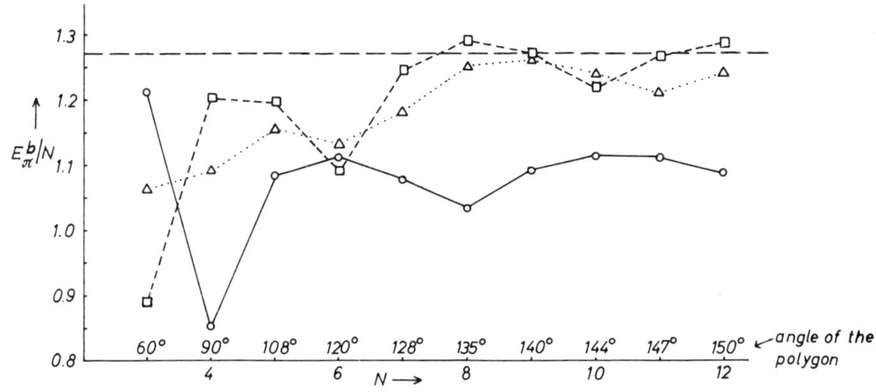

(——) = Hückel-type $(pd)_\pi$ system (radial d orbital)
(– – –) = Möbius-type $(pd)_\pi$ system (tangential d orbital)
(-------) = $(N+1)$ linear $(pd)_\pi$ -polyene-chain model (without cyclic conjugation)

Fig. 11. Specific π-bond energies (E^b_π/N) corrected for the deviations from optimum $(pd)_\pi$ overlap

Now the pattern has changed drastically relative to Fig. 9. Practically all the systems containing one d-orbital center are below the limit of aromaticity except the Möbius-type eight- and twelve-membered ring compounds, for which no examples of chemical compounds are known (compare Fig. 1: the phosphonitrilic compounds *16a* and *16b* are of the unfavourable Hückel type and already ruled out by consideration of the idealized systems).

Taking the most common cases, a monocyclic five-membered ring compound with one d-orbital center, the Möbius-type arrangement should be energetically favoured over the $(N+1)$ linear polyene-chain model, whereas for a monocyclic six-membered ring compound with one d-orbital center the $(N+1)$ linear polyene-chain model is more favourable than a Hückel-type arrangement, but both six-membered ring models yield lower values of specific energy than either five- or seven-membered ring systems.

2. Effects of Heteroatoms

The influence of a heteroatom on a π system may be expressed in HMO theory by a change of the atomic Coulomb integral parameter α and of the bond integral parameter β, both of which may be treated by use of perturbation theory [6,7]. As the change in β is caused by a change in the overlap integral, it may be treated together with corrections due to the actual geometry of the molecule. Generally, the change of α integrals by $\delta\alpha_\mu$ corresponds to a reduction of symmetry which causes a splitting of degenerate molecular-orbital energy levels. The corresponding second-order change in π-bond energy [117] is given by (12).

$$\delta E_\pi^b = (q_\mu - 1)\, \delta\alpha_\mu + \frac{\pi_{\mu\mu}}{2 \cdot \beta}\, \delta\alpha_\mu^2 \tag{12}$$

For an even cyclopolyene (alternating hydrocarbon) the π-electron density q_μ is equal to unity at each atom. Therefore, from (12), an increase in π-bond energy always follows the introduction of a heteroatom. But at the same time the cyclic delocalization energy, which is the difference in π-bond energy between the cyclic and the corresponding linear polyenes, will be decreased because the maximum atom self-polarizability of a linear polyene chain always has a higher value than that of the corresponding cyclopolyene [120].

For charged, odd-membered rings the change in π-bond energy will be determined mainly by the larger first-order term in (12). But the effect of an electronegative heteroatom in a negatively charged odd-membered ring will also be to increase specific π-bond energy with simultaneous reduction of cyclic delocalization energy, in other words, the introduction of a strongly electronegative heteroatom in a cyclopolyene can never lead to an increase in "aromaticity".

V. Predictions for $(pd)_\pi$ Systems Using Inner Orbitals

In transition metal complexes the magnitude of inner *d* orbitals, which is determined for Slater-type orbitals by the atomic screening constant, may be strongly affected by the bonding of coordinated atoms. As Clementi [108] has shown, the screening constant does not depend (as assumed in Slater's rules) only on the quantum numbers *n* and *l*, the nuclear charge, and the number of screening electrons, but also on the total number of electrons present (i.e. the number of electrons outside the shell in question) and for valence electrons on total angular momentum and the spin multiplicity of the atom.

For an octahedral complex the π-bonding leads in the (d^2sp^3) hybridization model to an increase in the number of electrons in the group of *d* orbitals which may be treated by Slater's rules, as well as to an increase of electrons in the next outer shell. As it is probable that the π-coordinated electrons may not be completely assigned to the transition metal ion, and as the more accurate tables of Clementi [108] contain only values for neutral atoms, the resulting effect can only be estimated by the use of Slater's rules. Table 2, p. 11, shows that the screening constant is increased; this leads to an increase in *d*-orbital size which is reflected in the magnitude of the overlap integral for a $N(2p_\pi)$-$Fe(3d_\pi)$ bond in the range 0.16 to 0.11. This crude estimate can only be improved by examining this effect by means of accurate calculations.

1. Transition Metal Chelates of 1,2-Diimines

For transition metal chelate rings of type *23* of Fig. 2, which contain 1,2-diimine ligands, the three known structurally different types of complexes, A, B, and C, have to be treated separately in a π-electron model:

Type A contains only one chelate ring and is formed in molybdenum (0)-carbonyl chelates [121] *41* or in copper(I) chelates [122] *42*, both synthesized and studied experimentally and theoretically by Bock and tom Dieck [121,123].

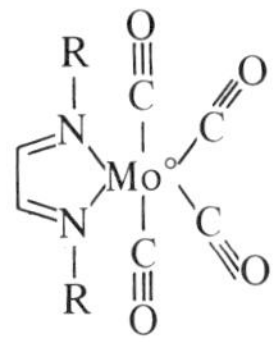

Type A: *41*

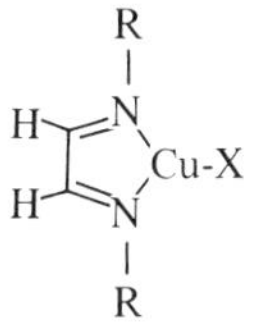

42

Type B, containing two chelate rings in planar quadratic arrangement, may be observed in 1,2-diimine chelates [124,125] *43*. But an X-ray structure determination of nickel glyoxime [126] showed that it does not possess ideal D_{2h} symmetry in the crystal because of contacts to atoms of neighbouring molecules.

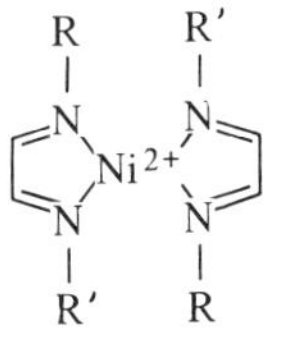

R = OH
R' = O⁻

Type B: *43*

Type C, with three chelate rings in an octahedral arrangement, is observed in Krumholz's [68] iron(II) complex *23*

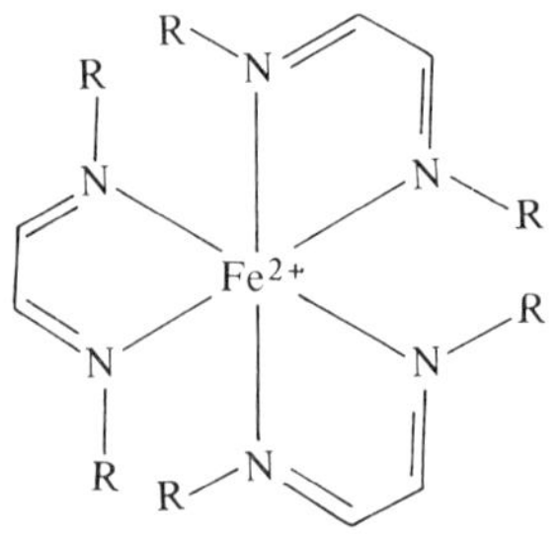

$R = CH_3$

Type C: *23*

2. Bonding Considerations

An accurate calculation transition metal chelates would have to take into account all orbitals up to the valence-electron shell having the proper symmetry to combine either to σ or to π bonds. But in approximate methods only the valence electrons are considered [127]. Usually the bonding in transition metal chelates like *23, 41, 42* or *43* is decribed by a back-donation model which considers only the interaction between the highest filled d orbitals of the metal and the empty ligand antibonding MO's [69,121,123] which are all of π symmetry. If one extends his model and considers the interaction with all bonding and antibonding π MO's and the d orbitals of π symmetry, one obtains the π-electron model which will be used here.

Octahedral and quadratic planar transition metal chelates have the geometrical peculiarity that the angle between σ bonds is found experimentally to be always very close to 90°. Therefore we do not have to take into account the reduction of overlap integrals for radial or tangential d-orbital orientations in dependence upon the polygon size, as shown in Fig. 10, but can deal with a constant geometry at the d-orbital center.

For type-A chelate rings *41* or *42* which belong to the C_{2v}-symmetry point group only, the d_{xz}, d_{yz} and p_z orbitals of the metal may take part in π-bonding if the coordinate axes are oriented as shown in *44*.

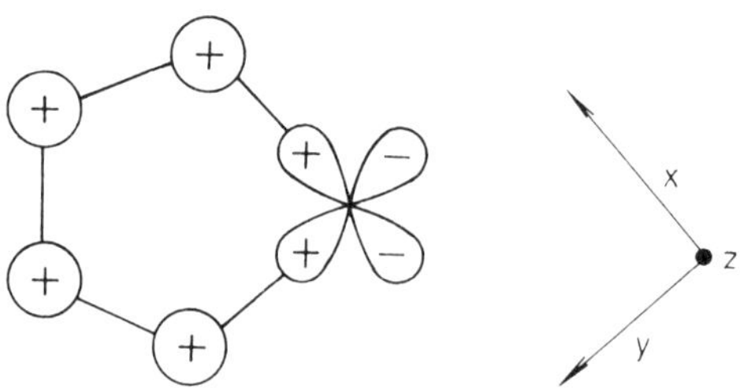

Bird's eye view in z-direction

44

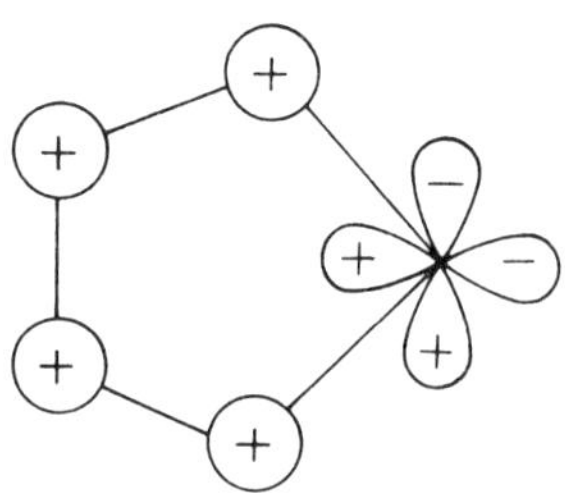

45

But as we assume the p_z orbital to be involved mainly in σ-bonding by means of d^2sp^3-hybrid orbitals to the axial carbonyl group, we consider only the two d_π orbitals. The σ bonds to the metal are formed by overlap of nitrogen lone-pair orbitals and two metal d^2sp^3 AO's. These may be pointing in the direction of the σ bonds as in *44* or in the direction shown in *45* which corresponds to the linear combination of both d orbitals (6).

The assumption of an angle of 90° between the σ bonds at the metal makes descriptions *44* and *45* equivalent. Model *44* has to be preferred if one constructs the π MO's from individual AO's. The $(pd)_\pi$-overlap integral is not reduced and it leads to the prediction of no cyclic conjugation across the metal. The resulting six-membered linear polyene-type π system is occupied by eight π electrons, i.e. two electrons enter antibonding MO's.

Model *45* may be used to describe the interaction of the two d_π orbitals with the four ligand π MO's but with reduced $(pd)_\pi$ overlap by cos 45°. The result are also six π MO's which are occupied by eight π electrons. For a full treatment, the interactions with the π systems of the ligands as well as those in the σ system have to be considered too.

In type-B chelate systems *43* with D_{2h} symmetry, a complete conjugation by use of two d orbitals is possible, leading to a ten-membered Hückel-type π system *46* which can be formed with an inversion of signs in one of the rings, yet without geometrical reduction of the magnitude of $(pd)_\pi$-overlap integrals.

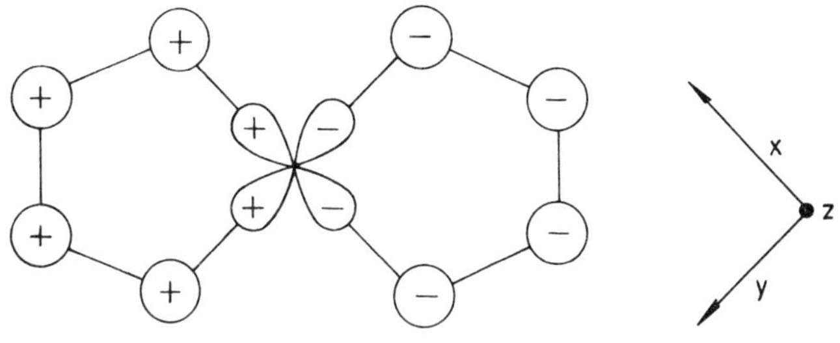

46

But this ten-membered Hückel-type $(pd)_\pi$ system, which would lead to an aromatic system if it were occupied by ten π electrons, has to take up twelve π electrons. The two additional π electrons in the anti-bonding MO's reduce the total stability so that, again, no high specific π-bond energy can be expected, although cyclic delocalization is possible. The idealized π model in Fig. 8 shows degeneracy for the lowest antibonding MO's but the two additional π electrons enter a nondegenerate MO because in D_{2h} symmetry of *46* only accidental degenracies are possible. (Of the ten MO's, three belong to the symmetry class B_{2g}, three to B_{3g}, both of which involve the d_{xz} and d_{yz}

orbitals, two belong to B_{1u}, and may form π bonds with the p_z orbital of the metal if this is considered in addition, and two belong to A_u, and are located only on the ligands without involvement of metal orbitals of π symmetry.)

A more advanced extended Hückel-type calculation for nickel glyoxime was performed by Ingraham [128] but he did not perform a population analysis [129] to show the extent of d-orbital participation.

Octahedral chelate systems of type C with D_3 symmetry *23* form by use of the three d_π orbitals (d_{xy}, d_{xz} and d_{yz}) a cyclically delocalized π system which connects all three chelate rings, as shown in *47*.

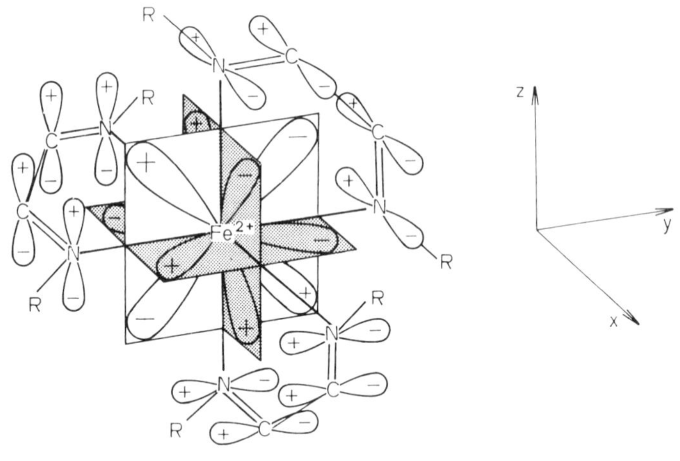

47

Now it is necessary to use the possibiliy for perpendicular π-bonding of d orbitals shown on pp. 19 and 20. The result is one agglomerated, cyclically conjugated π system comprising fifteen atomic orbitals of π symmetry which contain one nodal plane, i.e. of the Möbius type. This would show aromatic behaviour if filled with sixteen π electrons but, here again, two additional electrons have to be placed in an antibonding MO which is nondegenerate because of the D_3 symmetry of the molecule. Therefore no special aromatic stability is to be expected. The three p orbitals (p_x, p_y and p_z) of the metal would have the proper symmetry to form three independent π systems with five membered rings (each p orbital having one chelate ring), but they would also point directly towards the coordinated ligands, giving rise to strong σ-bonding.

3. Comparison of Chelate Rings of Different Ring Sizes

The π-electron model may be used to predict relative stabilities for chelate rings of different ring sizes, i.e. for one row of Fig. 2 (for the same type of coordinating group, as well as for the same transition metal ion which is supposed to be able to provide six π electrons in three *d* orbitals). For each chelate ring size, the structural types A, B or C may be obtained, having C_{2v}, D_{2h} and D_3 symmetry, respectively. Type A will always preferentially form a linear $(N+1)$ polyene-type system, for which practically no cyclic delocalization is predicted. Type B will form a cyclic conjugated Hückel-type π system and type C a Möbius-type π system. In all of these chelate types at least one antibonding MO will be occupied in addition. This is not so bad as it sounds because, in hetero-substituted compounds, the gap between bonding and antibonding MO's is not as big as in unsaturated hydrocarbons, for example, and in any case it decreases with increasing ring size. Therefore, in this π-electron approach, the ground-state stability will be determined both by the MO-pattern of the basically formed π system and to a smaller extent by the π-electron occupation number.

For four-membered chelate rings and for six-membered rings of type B, Hückel-type π systems are formed with 8 and 12 MO's, respectively, corresponding to unfavourable $(4m)$ Hückel $(pd)_\pi$ systems.

For four-membered chelate rings of type C, the Möbius-type $(pd)_\pi$ systems formed have the favourable twelve-membered ring into which eighteen π electrons have to be fitted, with three antibonding MO's occupied, whereas the six-membered chelate ring of type C, the Mason-type system, contains eighteen MO's into which twenty-four π electrons have to be placed, i.e. two antibonding MO's are filled.

Therefore, it is predicted that the four- or six-membered chelate rings of types B and C will be less stable than the corresponding five-membered chelate ring system.

Among the seven-membered rings, only type B with a fourteen-membered Hückel-type $(pd)_\pi$ system occupied by sixteen π electrons will be favourable.

Type C (Möbius-type twenty-one-membered) has to be occupied by twenty-four π electrons (2 anti-bonding MO's filled).

The stability of type-A chelates which show no cyclic delocalization should increase with increasing chelate ring size.

VI. Conclusion

In most of the cyclic $(pd)_\pi$ systems which may be formed by use of either outer or inner *d* orbitals, cyclic delocalization through the *d*-orbital center is possible and leads to an increase in stability, but in no case will the $(pd)_\pi$ system alone lead to an extraordinary ground-state stabilization which can be classified as aromatic.

VII. References

1) Lloyd, D.: Carbocyclic Non-Benzenoid Compounds, p. 11. Amsterdam: Elsevier 1966.
2) Jones, A. J.: Rev. Pure and Appl. Chem. (Melbourne) *18,* 253 (1968).
3) Baker, G. M.: Aromaticity and Aromatic Character, p. 36. Cambridge University Press 1969.
4) Snyder, J. P.: Nonbenzenoid Aromatics, Chap. 1. New York: Academic Press 1969.
5) Hückel, E.: Z. Physik *70,* 204 (1931); *72,* 310 (1931); *76,* 628 (1932); Z. Elektrochem. *43,* 752 (1937).
6) Streitwieser, A., Jr.: Molecular Orbital Theory for Organic Chemists. New York: J. Wiley & Sons. 1961.
7) Heilbronner, E., Bock, H.: Das HMO-Modell und seine Anwendung, Bd. 1–3. Weinheim/Bergstr.: Verlag Chemie 1968, 1970.
8) Craig, D. P., Maccoll, A., Nyholm, R. S., Orgel, L. E., Sutton, L. E.: J. Chem. Soc. *1954,* 332.
9) Mitchell, K. A. R.: Chem. Rev. *69,* 157 (1969).
10) Coulson, C. A.: Nature *221,* 1106 (1969).
11) Clementi, E., Raimondi, D. L., Reinhardt, W. P.: IBM Res. Rept. RJ – 431 (1962).
12) Quin, L. D., Bryson, J. G., Moreland, C. G.: J. Am. Chem. Soc. *91,* 3308 (1969).
13) Brown, D. A.: J. Chem. Soc. *1962,* 929.
14) Märkl, G.: Angew. Chem. *78,* 907 (1966).
15) Bart, J. C. J., Daly, J. J.: Angew. Chem. *80,* 843 (1968).
16) Märkl, G.: Angew. Chem. *75,* 669 (1963).
17) Daly, J. J., Märkl, G.: Chem. Commun. *1969,* 1057.
18) Thewalt, U.: Angew. Chem. *81,* 783 (1969).
19) – Bugg, C. E., Hettche, A.: Angew. Chem. *82,* 933 (1970).
20) Märkl, G.: Z. Naturforsch. *18b,* 1136 (1963); Angew. Chem. *77,* 1109 (1965).
21) – Schubert, H.: Tetrahedron Letters *1970,* 1273.
22) Mason, S. F.: Nature *205,* 495 (1965).
23) Longuet-Higgins, H. C.: Trans Faraday Soc. *45,* 173 (1949).
24) Sappenfield, D. S., Kreevoy, M.: Tetrahedron *19,* Suppl. 2, 158 (1963).
25) Bielefeld, M. J., Fitts, D. D.: J. Am. Chem. Soc. *88,* 4804 (1966).
26) Clark, D. T.: Tetrahedron *24,* 2663 (1968).
27) Julg, A., Bonnet, M., Ozias, Y.: Theoret. Chim. Acta *17,* 49 (1970).
28) Dobyns, V., Pierce, L.: J. Am. Chem. Soc. *85,* 3553 (1963). – Momany, F. A. Bonham, R. A.: J. Am. Chem. Soc. *86,* 162 (1963).
29) Bak, B., Nygaard, L., Pedersen, G. J., Rastrup-Andersen, J.: J. Mol. Spectr. *19,* 283 (1966).
30) Blackman, G. L., Brown, R. D., Burden, F. R., Kent, J. E.: Chem. Phys. Letters *1,* 379 (1967).
31) Levine, D. M., Krugh, W. D., Gold, L. P.: J. Mol. Spectr. *30,* 459 (1969).
32) Zahradník, R., Koutecky, J.: Collection Czech. Chem. Commun. *26,* 156 (1961).
33) Ray, N. K., Narasimhan, P. T.: Theoret. Chim. Acta *5,* 401 (1966).
34) Markov, P., Skancke, P. N.: Acta Chem. Scand. *22,* 2051 (1968).
35) Howell, P. A., Curtis, R. M., Lipscomb, W. N.: Acta Cryst. *7,* 498 (1954).
36) Degani, I., Fochi, R., Vincenzi, C.: Tetrahedron Letters *1963,* 1167.
37) Molanaar, E., Strating, J.: Tetrahedron Letters *1965,* 2941.
38) Degani, I., Fochi, R., Vincenzi, C.: Gazz. *94,* 203 (1964).
39) Fabian, J., Mehlhorn, A., Zahradník, R.: Theoret. Chim. Acta *12,* 247 (1968).
40) Hess, H., Forst, D.: Z. Anorg. Allgem. Chem. *342,* 240 (1966).

41) Dougill, M. W.: J. Chem. Soc. *1963*, 3211.
42) Wilson, A., Carroll, D. F.: J. Chem. Soc. *1960*, 2548.
43) Wiegers, G. A., Vos, A.: Proc. Chem. Soc. *1962*, 387.
44) Mc Geachin, H. D., Tromans, F. R.: J. Chem. Soc. *1961*, 4777.
45) Schlueter, A. W., Jacobson, R. A.: J. Am. Chem. Soc. *88*, 2051 (1966).
46) Hazekamp, R., Migchelsen, T., Vos, A.: Acta Cryst. *15*, 539 (1962).
47) Wagner, A. J., Vos, A., de Boer, J. L., Wichertjes, T.: Acta Cryst. *16*, A 39 (1963).
48) Wiegers, G. A., Vos, A.: Acta Cryst. *14*, 562 (1961); *16*, 152 (1963).
49) Wagner, A. J., Vos, A.: Rec. Trav. Chim. *84*, 603 (1965).
50) Paddock, N. L., Trotter, J., Whitlow, S. H.: J. Chem. Soc. A, *1968*, 2227.
51) Glemser, O.: Angew. Chem. *75*, 697 (1963).
52) Paddock, N. L.: Quart. Rev. *18*, 168 (1964).
53) Craig, D. P.: Chem. Soc. Special Publ. No. 12, p. 343. Bristol Sympos. London 1958.
54) – Paddock, N. L.: Nature *181*, 1052 (1958).
55) – J. Chem. Soc. *1959*, 997.
56) – Heffernan, M. L., Mason, R., Paddock, N. L.: J. Chem. Soc. *1961*, 1376.
57) – Paddock, N. L.: J. Chem. Soc. *1962*, 4118.
58) – Mitchell, K. A. R.: J. Chem. Soc. *1965*, 4682.
59) Dewar, M. J. S., Lucken, E. A. C., Whitehead, M. A.: J. Chem. Soc. *1960*, 2423.
60) Bradley, W., Wright, I.: J. Chem. Soc. *1956*, 640.
61) Dwyer, I. P., Mellor, D. P.: J. Am. Chem. Soc. *63*, 81 (1941).
62) Barclay, G. A., Kennard, C. H. L.: J. Chem. Soc. *1961*, 5244.
63) – – J. Chem. Soc. *1961*, 3289.
64) Thorn, G. D., Ludwig, R. A.: The Dithiocarbomates and Related Compounds. Amsterdam: Elsevier Monographs 1962.
65) Bonamico, M., Dessy, G., Mariani, C., Vaciago, A., Zambonelli, L.: Acta Cryst. *19*, 619 (1965).
66) – – Mugnoli, A., Vaciago, A., Zambonelli, L.: Acta Cryst. *19*, 886 (1965).
67) – – – – – Acta Cryst. *19*, 898 (1965).
68) Krumholz, P.: J. Am. Chem. Soc. *75*, 2163 (1953).
69) – Iron(II)Diimine and Related Complexes. In: Structure and Bonding, Vol. 9, p. 139. Berlin-Heidelberg-New York: Springer 1971.
70) Figgins, P. E., Busch, D. H.: J. Am. Chem. Soc. *82*, 820 (1960).
71) Bayer, E.: Angew. Chem. *73*, 533 (1961).
72) – Fiedler, H., Hock, K.-L., Otterbach, D., Schenk, G., Voelter, W.: Angew. Chem. *76*, 76 (1964).
73) Schrauzer, G. N.: Advan. Organometal. Chem. *2*, 17 (1964).
74) – Accounts of Chem. Res. *2*, 72 (1969).
75) – Transition Metal Chem. *4*, 299 (1968).
76) McCleverty, J. A.: Progr. Inorg. Chem. *10*, 49 (1968).
77) Schrauzer, G. N., Mayweg, V. P.: J. Am. Chem. Soc. *87*, 1483 (1965).
78) Häfelinger, G.: Dissertation Univ. Tübingen 1965.
79) Daltrozzo, E., Feldmann, K.: Angew. Chem. *79*, 153 (1967); Ber. Bunsenges. Phys. Chem. *72*, 1140 (1969).
80) Fackler, J. P., Jr.: Progr. Inorg. Chem. *7*, 374 (1966).
81) Armit, J. W., Robinson, R.: J. Chem. Soc. *127*, 1604 (1925).
82) Calvin, M., Wilson, K. W.: J. Am. Chem. Soc. *67*, 2003 (1945).
83) Kuhr, M., Musso, H.: Angew. Chem. *81*, 150 (1969).
84) Bock, B., Flatau, K., Junge, H., Kuhr, M., Musso, H.: Angew. Chem. *83*, 239 (1971).
85) Barraclough, C. G., Martin, R. L., Stewart, I. M.: Austr. J. Chem. *22*, 891 (1969).
86) Witte, E. G.: Dissertation Univ. Tübingen.

87) Lennard-Jones, J. E.: Trans. Faraday Soc. *25,* 668 (1929).
88) Mulliken, R. S.: J. Chem. Phys. *3,* 375 (1935).
89) Lennard-Jones, J. E.: Proc. Roy. Soc. (London) A *198,* 1 *(1949);* A *198,* 14 *(1949).*
90) Pople, J. A.: Quart. Rev. *11,* 273 (1957).
91) Kutzelnigg, W., del Re, G., Berthier, G.: σ and π-electrons in Theoretical Organic Chemistry. In: Topics in Current Chemistry, Vol. 22. Berlin-Heidelberg-New York: Springer 1971.
92) Sinanoǧlu, O., Wiberg, K. B.: Sigma Molecular Orbital Theory, p. 211. New Haven and London: Yale University Press 1970.
93) Coulson, C. A.: Valence, 2nd ed., p. 241. London: Oxford University Press 1961.
94) Roothaan, C. C. J.: Rev. Mod. Phys. *23,* 69 (1951).
95) Klopman, G., O'Leary, B.: All-Valence Electrons S.C.F. Calculations of Large Organic Molecules. In: Topics in Current Chemistry, Vol. 15, p. 445. Berlin-Heidelberg-New York: Springer 1970.
96) Richards, W. G., Horsley, J. A.: Ab initio Molecular Orbital Calculations for Chemists. Oxford: Oxford University Press 1971.
97) Pople, J. A., Beveridge, D. L.: Approximate Molecular Orbital Theory. New York: McGraw-Hill 1970.
98) – Santry, D. P., Segal, G. A.: J. Chem. Phys. *43,* 5129 (1965).
99) – Segal, G. A.: J. Chem. Phys. *44,* 3289 (1966).
100) – Beveridge, D. L., Dobosh, P. A.: J. Chem. Phys. *47,* 2026 (1967).
101) Dewar, M. J. S., Klopman, G.: J. Am. Chem. Soc. *89,* 3089 (1967).
102) Hoffmann, R.: J. Chem. Phys. *39,* 1397 (1963).
103) Salem, L.: The Molecular Orbital Theory of Conjugated Systems. New York: W. A. Benjamin 1966.
104) Klessinger, M.: Mehrelektronenmodelle der organischen Chemie, Topics in Current Chemistry, Vol. 9, p. 400. Berlin-Heidelberg-New York: Springer 1968.
105) Pariser, R., Parr, R. G.: J. Chem. Phys. *21,* 466, 767 (1953).
106) Pople, J. A.: Trans. Faraday Soc. *49,* 1375 (1953).
107) Slater, J. C.: Phys. Rev. *36,* 87 (1930).
108) Clementi, E., Raimondi, D. L.: J. Chem. Phys. *38,* 2686 (1963).
109) Mulliken, R. S.: J. Am. Chem. Soc. *72,* 4493 (1950).
110) Häfelinger, G.: Tetrahedron *26,* 2469 (1970); *27,* 1635 (1971); Chem. Ber. *103,* 2902, 2922, 2941 (1970).
111) Mulliken, R. S., Rieke, C. A., Orloff, D., Orloff, H.: J. Chem. Phys. *17,* 1248 (1949).
112) Jaffé, H. H.: J. Chem. Phys. *21,* 258 (1953).
113) Häfelinger, G.: Tetrahedron Letters *1971,* 541; Tetrahedron *27,* 4609 (1971).
114) Zahradník, R.: Angew. Chem. *77,* 1097 (1965).
115) Ashe, A. J.: Tetrahedron Letters *1968,* 359.
116) Heilbronner, E.: Tetrahedron Letters *1964,* 1923.
117) Coulson, C. A.: Proc. Roy. Soc. (London) A *164,* 383 (1938).
118) Frost, A. A., Musulin, B.: J. Chem. Phys. *21,* 572 (1953).
119) Coulson, C. A., Longuet-Higgins, H. C.: Proc. Roy. Soc. (London) A *191,* 39 (1947); A *193,* 447 (1948).
120) Heilbronner, E., Straub, H.: HMO-Hückel Molecular Orbitals. Berlin-Heidelberg-New York: Springer 1966.
121) Bock, H., tom Dieck, H.: Chem. Ber. *100,* 228 (1967).
122) tom Dieck, H., Renk, I. W.: Chem. Ber. *100,* 228 (1967).
123) – – Chem. Ber. *104,* 110 (1971).
124) Tschugaeff, L. A.: J. Prakt. Chem. *75,* 153 (1907).
125) Pfeiffer, P., Buchholz, E.: J. Prakt. Chem. *124,* 133 (1930).

126) Callèri, M., Ferras, G., Viterbo, D.: Acta Cryst. *22,* 468 (1967).
127) Ballhausen, C. J.: Introduction to Ligand Field Theory. New York: McGraw-Hill 1962.
128) Ingraham, L. I.: Acta Chem. Scand. *20,* 283 (1966).
129) Mulliken, R. S.: J. Chem. Phys. *23,* 1833, 1841 (1955).

Received August 16, 1971

Organic Synthesis by Means of Transition Metal Complexes

Some General Patterns

Jiro Tsuji

Basic Research Laboratory, Toray Industries Inc., Kamakura, Japan

Contents

I. Introduction

The art of organic synthesis has been enriched by the development of new methods: remarkable advances have been made in the last 15 years by the use of transition metal complexes. Many interesting syntheses, which would have been impossible by conventional organic methods, have been achieved. Transition metal compounds, unlike compounds of non-transition metals such as Mg, Li, or Zn, have several characteristic properties which contribute to their usefulness in organic synthesis:

Transition metal compounds have a strong affinity for such substrates as carbon monoxide, olefins, acetylenes, hydrogen, and other simple unsaturated compounds, and activate them by forming complexes. *Complex formation* is indispensable for bringing these compounds and other organic substances into reactions. The transition metals have the ability to stabilize, through coordination, a wide variety of species as σ-bonded or π-bonded ligands.

In addition, transition metal compounds have the ability to donate additional electrons or accept electrons from organic substrates and can change both their valence and their coordination number reversibly. These properties play an important role in organic synthesis, especially in *catalytic processes.* The ability to serve as catalysts in organic reactions is the most important property of the transition metal compounds. Reaction mechanisms involving intermediate organic structures, which are prohibitively endothermic in the absence of transition metal catalysts, are made feasible in their presence.

The mechanisms of the usual organic reactions are now clearly established, and the reactions are classified as ionic, radical, and molecular. More detailed classifications have also been made. The mechanisms of many reactions involving non-transition metal compounds are clear enough: for example, in the Grignard or Reformatsky reaction, the first step is the irreversible oxidative addition of alkyl halides to form Mg-carbon or Zn-carbon bonds, in which the carbon is considered to be a nucleophilic center or carbanion which reacts with various electrophiles.

The mechanisms of the organic reactions involving transition metal complexes are, however, not completely clear; certainly, the reactions proceed through the formation of carbon-metal σ-bonds but the chemical properties of these bonds are somewhat obscure.

This review seeks to establish some *general patterns of organic reactions using transition metal complexes* and hence to reach some understanding of them. In other words, it discusses how σ-bonds are formed from simple molecules and metal complexes, and how they are then transformed into organic compounds, and gives typical examples. There are many types of reactions using transition metal complexes and it is impossible to survey them all, therefore only typical patterns are discussed. The references are by no means exhaustive because a complete literature survey is not the purpose of this review. The typical examples included are mostly reactions of *noble metal complexes,* chosen because of the author's own interests. Many synthetic reactions involving transition metal complexes can be understood as a succession of simple, fundamental reactions. These simple reactions are surveyed here.

The first essential step in the organic reactions of transition metal compounds is the formation of metal σ-bonds, either directly or indirectly, with hydrogen, carbon, oxygen, nitrogen, and halogen. There are several ways in which σ-bonds may form between metal complexes and simple molecules [1]. These are surveyed first.

II. σ-Bond Formation

1. Oxidative Addition Reactions

This is a well-established class of reaction and several reviews have been published [2,3]. The reaction is widely observed with transition metal complexes, especially with noble metal complexes, and is most important for the direct formation of σ-bonds from metal complexes and simple molecules. The reaction can be expressed by the following scheme:

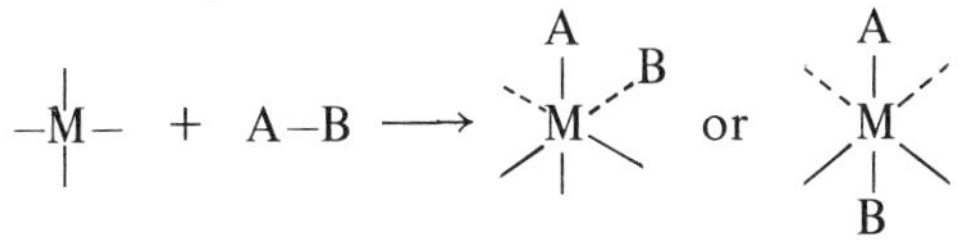

This reaction is an essential step in a wide variety of catalytic processes.

It involves the addition of a covalent molecule to the metal with cleavage of a covalent bond and can be considered a two-electron oxidation of the metal. One of the necessary conditions for this reaction is that the metal complexes to be oxidized should be coordinatively unsaturated. With the transition metal complexes, a saturated coordination number is determined by the configuration of the metal *d*-electrons (d^n). Thus, for d^6 configuration metals, six-coordination is regarded as saturated, and similarly, five-coordination for d^8 metals, and four-coordination for d^{10} metals. If the complexes with these d^n have less than these coordination numbers, they are said to be coordinatively unsaturated, and oxidative addition reaction may occur. In other words, coordinative unsaturation means that there are vacant sites on the complexes.

Coordinative unsaturation includes potential, or solvent-occupied, vacant sites. For examples, with four-coordinate $RhCl(PPh_3)_3$, Rh has the d^8 configuration and the complex is apparently unsaturated. Furthermore, in solution, one mole of triphenylphosphine dissociates and hence the complex becomes highly unsaturated [4], making oxidative addition feasible. This property partly explains the usefulness of this complex as a catalyst for various reactions.

On the other hand, with saturated complexes, a vacant site must be created if the metal is to take part in oxidative addition reactions. For example, $Fe(CO)_5$ is a d^8 complex and the five-coordination is saturated and stable. In the reaction with $Fe(CO)_5$, one of the carbon monoxide ligands must be liberated to make the complex unsaturated and bring it to an excited state. For this purpose, radiation or heat is necessary as an energy source for the initiation of the reaction using $Fe(CO)_5$ [5].

$$Fe(CO)_5 \longrightarrow Fe(CO)_4 + CO$$

The behavior of saturated $Ni(CO)_4$ (d^{10}) as a catalyst can be understood in a similar way.

Some saturated complexes are easily converted into unsaturated states without an extra energy source. For example, four-coordinate Pt and Pd phosphine complexes, [$M(PPh_3)_4$, M = Pt or Pd] have d^{10} configurations and are regarded as saturated. However, it was proved by Malatesta and Cariello that the coordinated phosphines dissociate in solution, forming dicoordinate complexes which readily undergo oxidative addition reactions, as will be shown later [6,7]. These complexes are said to be potentially unsaturated, and are thus useful as catalysts in several reactions.

$$Pd(PPh_3)_4 \longrightarrow Pd(PPh_3)_2 + 2PPh_3$$

The usefulness of transition metal complexes for many organic synthesis is a consequence of their coordinative unsaturation and the ease with which they undergo oxidative addition. This ease is determined by the nature of both the metals and their ligands. In addition, the stability of the products of the oxidative addition is important in enabling further reactions to occur. If the product of the oxidative addition is too stable and can easily be isolated as a stable compound, then it is not useful for organic synthesis. For example, $Ir(CO)Cl(PPh_3)_2$ undergoes many oxidative additions and the products can be isolated easily, but it is not very useful for organic synthesis [8]. With $RhCl(PPh_3)_3$, the product of the oxidative addition is a five-coordinate d^6 complex and still unsaturated. Thus, it may undergo further transformation to form a six-coordinate d^6 complex.

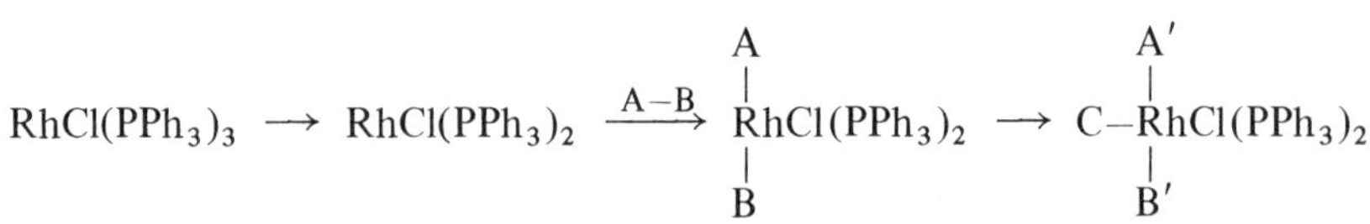

The most often cited compounds which add to metal complexes oxidatively are: H_2, HX, RCOX (X = halogen), R_3SiH and RSO_2X. Further examples from the recent literature are surveyed in the following.

The first example is the *activation of hydrogen,* attached to various other atoms, by the formation of metal hydrides. Hydrogen abstraction from an activated carbon-hydrogen bond is one of the most interesting reactions. Indeed, the carbon-hydrogen bond is ubiquitous in organic chemistry and the unusual ability of transition metal catalysts to assist in its breaking and making is particularly useful. Many compounds which have active hydrogens are capable of adding to metal complexes.

Oxidative addition of aldehydes is expected from the mechanism of their decarbonylation reactions, catalyzed by rhodium and palladium catalysts [9,10]. Harvie and Kemmitt reported the formation of the following diacyl complex by the reactions of the aldehydes with $Pt(PPh_3)_4$ [11].

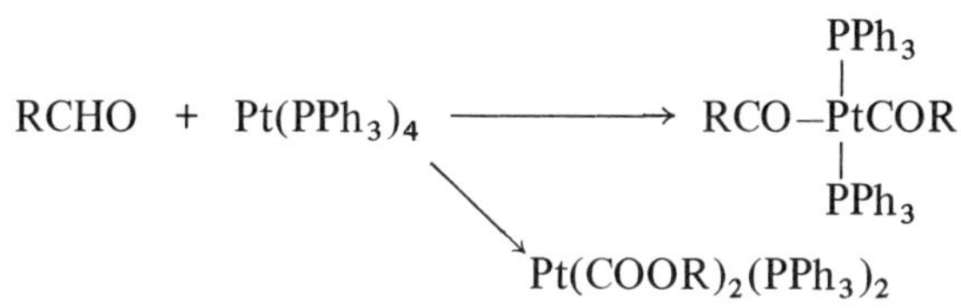

However, Tripathy and Roundhill proposed that the product of the reaction was not the diacyl complex but dicarboxylate [12], and further confirmation of the reactions is necessary.

Another example of reactions with carbon-hydrogen bond cleavage is shown by acetylene complexes [13,13a].

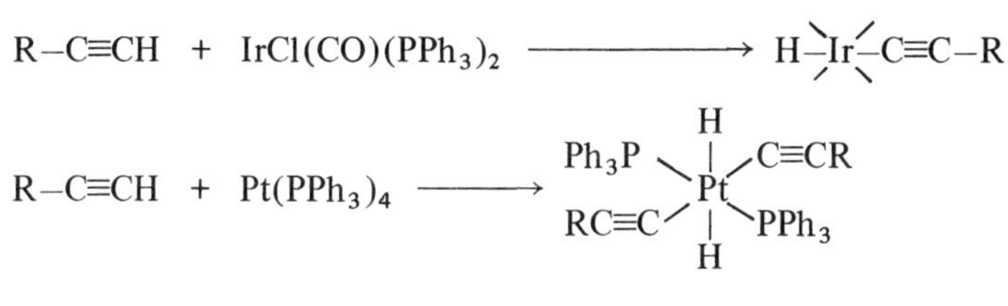

Carbon-hydrogen bonds in aromatic compounds can be split when they are suitably located [109]. For example, in the following reaction a four-membered ring is formed by the replacement of one of the phenyl hydrogens of the triphenylphosphine in the complex by iridium through C-H bond splitting [14]. Similarly a five-membered ring is formed by cleavage of a C-H bond in the benzene ring of a triphenylphosphite complex with formation of a metal carbon bond [15,16].

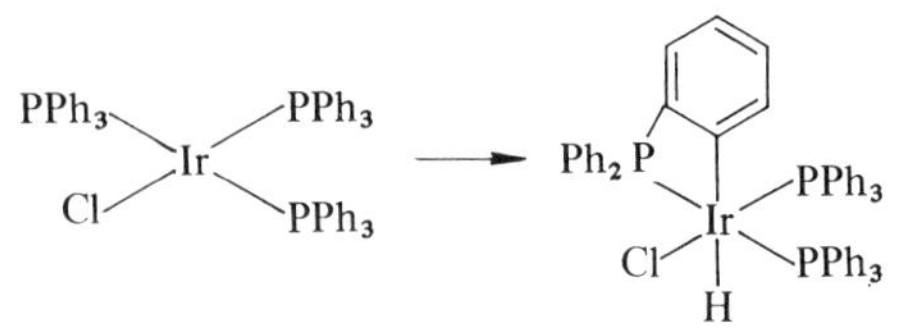

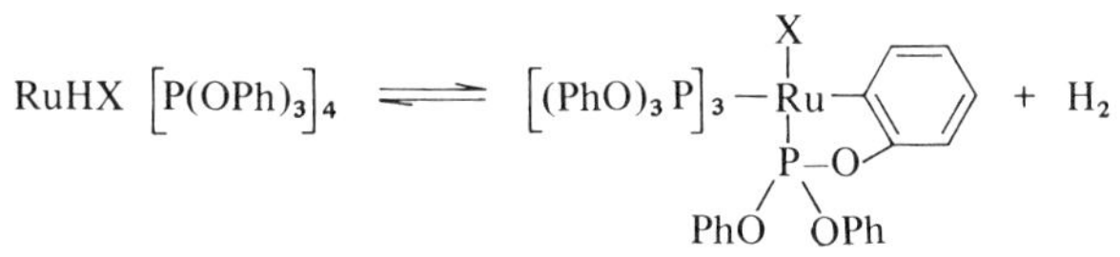

Formation of a rhodium tetraphenylporphine complex having a Rh-phenyl σ-bond was reported by Fleischer and Lavalles [17].

The following intramolecular isomerization is an example of the splitting of a saturated C-H bond [18].

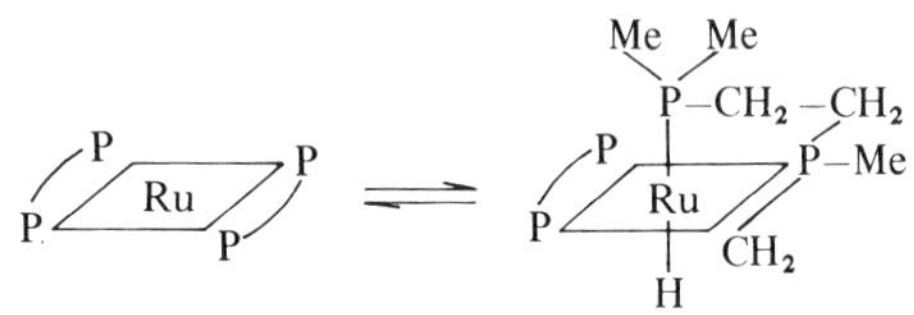

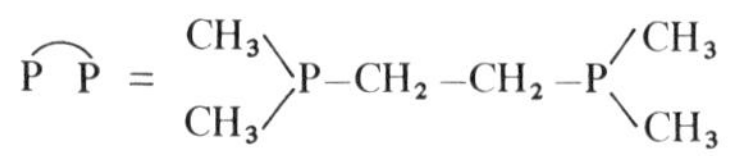

In the above four-coordinate ruthenium complex, the ruthenium has the d^8 configuration and is unsaturated, and hence tends to form a saturated six-coordinate d^6 complex. This tendency is certainly the driving force of the C-H bond splitting. These C-H bond splittings show interesting possibilities from the standpoint of organic synthesis.

In related reactions, C-H bond cleavage takes place in *oligomerization* and cooligomerization of various olefinic compounds to form unsaturated acyclic compounds. In the mechanism of these addition reactions, hydrogen transfer is an important step. This process certainly proceeds through C-H bond cleavage to give a metal-H bond as an intermediate. Hydrogen then migrates intermolecularly to another molecule. For example, in the dimerization of deuterated butadiene with a cobalt complex, the following product was obtained, clearly showing the intermolecular hydrogen shift [19].

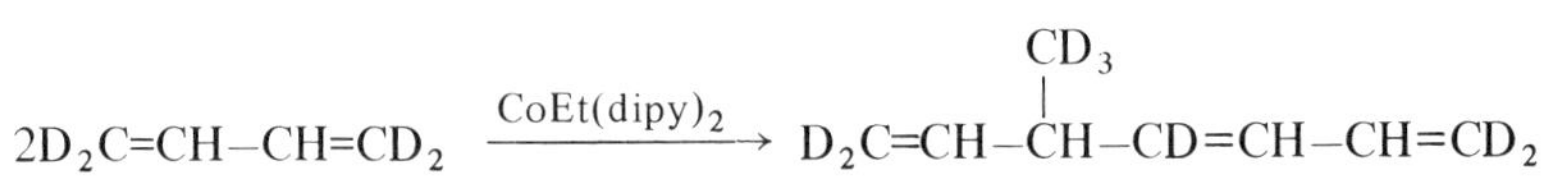

The N-H bond is also capable of undergoing oxidative addition. Roundhill found that amines and imines undergo addition reaction with platinum complexes to form metal-nitrogen bonds [20].

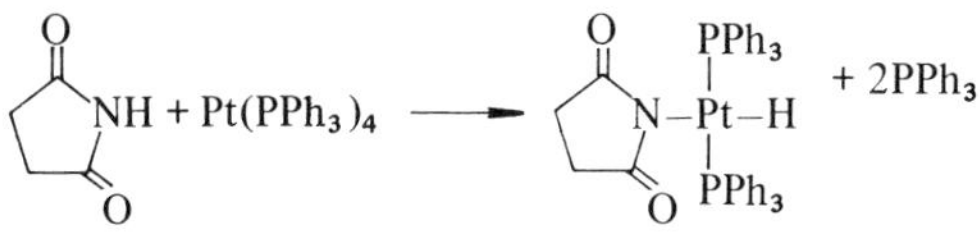

$$R_2NH + Pt(PPh_3)_4 \longrightarrow R_2N\text{-}Pt(PPh_3)_2\text{-}H + 2PPh_3$$

A similar addition product was formed in the reaction of *trichlorosilane* with the same platinum complex [21].

$$Cl_3SiH + Pt(PPh_3)_4 \longrightarrow Pt(SiCl_3)(PPh_3)_2 + H_2 + 2PPh_3$$

Although *vinyl halogens* are regarded as inert in organic reactions, Fitton and McKeon found that they can also add to $Pd(PPh_3)_4$ with cleavage of the carbon-chlorine bond [22,23].

$$CH_2{=}CCl_2 + Pd(PPh_3)_4 \longrightarrow CH_2{=}C(Cl)\text{-}PdCl(PPh_3)_2 + 2PPh_3$$

As explained before, these four-coordinate Pt and Pd phosphine complexes are potentially unsaturated.

Furthermore although there is no definite evidence, the oxidative additions of active hydrogen compounds, such as alcohols, water, hydrogen cyanide, and active methylene compounds, are reasonably predictable from their reactivity. Some reactions involving these compounds can be explained by oxidative addition with splitting of bonds to hydrogen. Several examples of this type are surveyed later.

Not only noble metal complexes, but also *nickel complexes* undergo oxidative addition reactions. Fahey found that a variety of vinyl and aryl halides react with $(R_3P)_2Ni(C_2H_4)$ to form a stable carbon-metal σ-bond [24]. For example, tetrachloroethylene affords *trans*-chloro(trichlorovinyl)*bis*(triphenylphosphine)nickel.

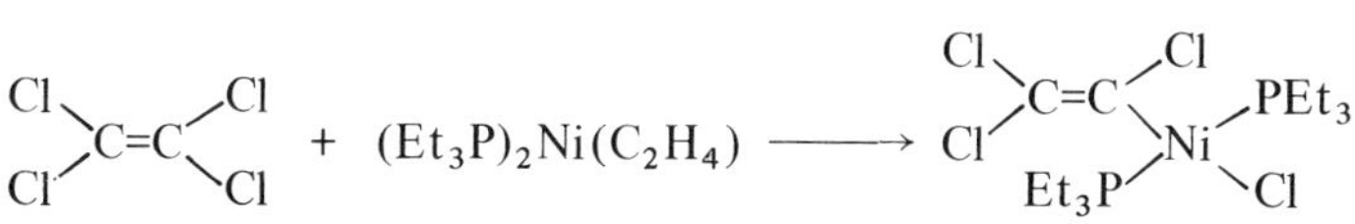

Jones and Wilke found that the following compounds which have active hydrogens undergo addition reaction with zerovalent nickel complexes [25].

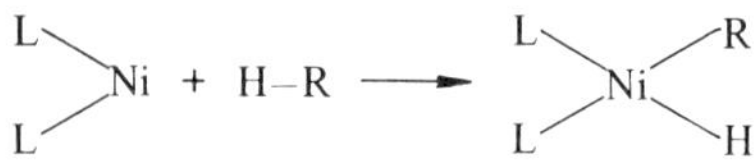

HR: HCl, CH_3COOH 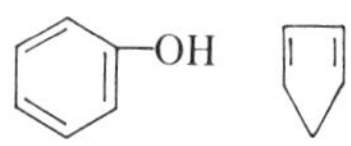

Carbon-carbon bond splitting would be a most interesting reaction from the standpoint of organic synthesis. But, so far, only one example of such a reaction is known. Schott and Wilke reported that hexaphenylethane undergoes a carbon-carbon bond cleavage in the following way [26]. However, this is possible only due to a large steric effect and seems to be an exception.

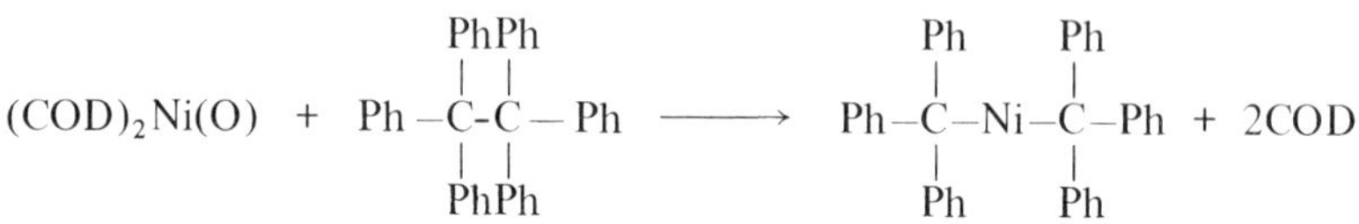

COD: cyclooctadiene

Examples of C-O and O-O bond splitting are also known

$$\text{R–O–O–R} + \text{Ni(O)} \longrightarrow \text{R–O–Ni–O–R}$$

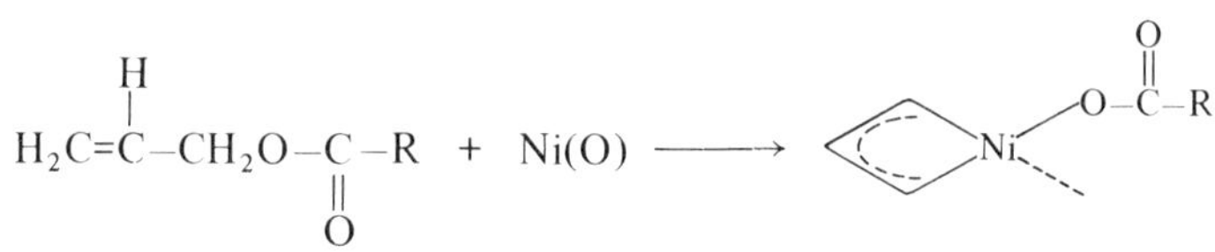

Reactions of *halogen-containing compounds* with $Ni(CO)_4$, $Fe(CO)_5$, or $Fe_3(CO)_{12}$ are widely observed. In these reactions, the formation of intermediate σ-alkyl complexes by the addition reaction, followed by carbon monoxide insertion to form acyl complexes, is assumed. From these acyl complexes, many useful carbonyl compounds can be synthesized. Compounds having relatively active halogens such as

$$\text{C=C–C–X}, \quad \text{–C(=O)–C–X}, \quad \text{NC–C–X}, \quad C_6H_5CH_2\text{–X}, \text{ or } C_6H_5\text{–I}$$

take part in the reactions. Some examples are given below.

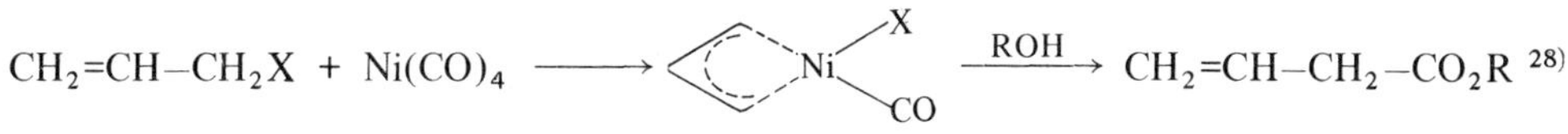

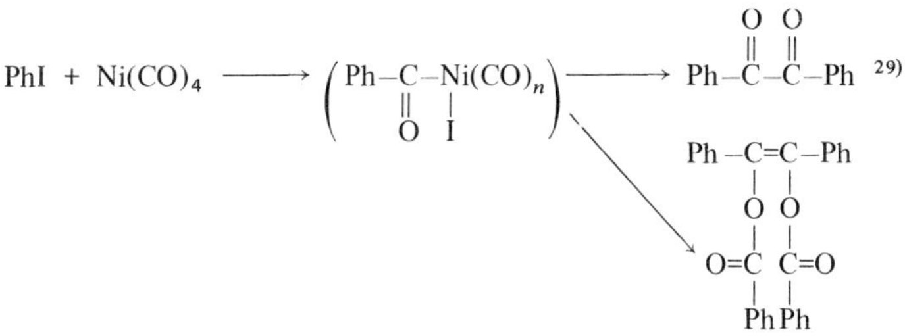

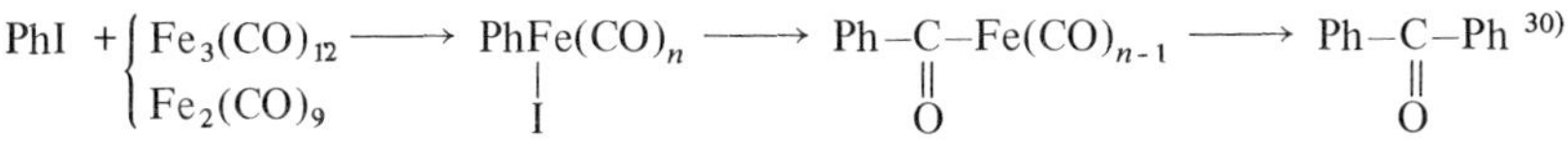

A further example of this type of reaction was shown by Tsutsumi and Yoshisato in *furane formation*. α-Bromoketones react with $Ni(CO)_4$; the products of the reaction are different depending on the solvent used in the reaction. Thus, in tetrahydrofurane, diketone is obtained. On the other hand, furane derivatives are obtained in dimethylformamide. As an intermediate of the furane formation, an epoxy compound was isolated.

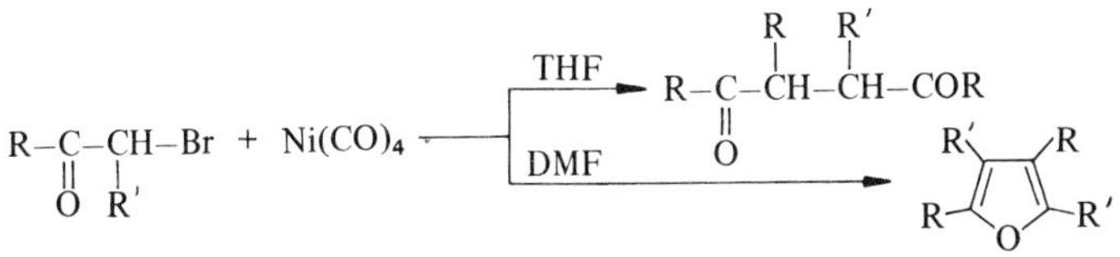

The reaction can be explained by oxidative addition of the carbon-bromine bond to form a nickel-carbon bond, followed by a nucleophilic attack of the nickel-carbon bond on the carbonyl of the other bromoketone.

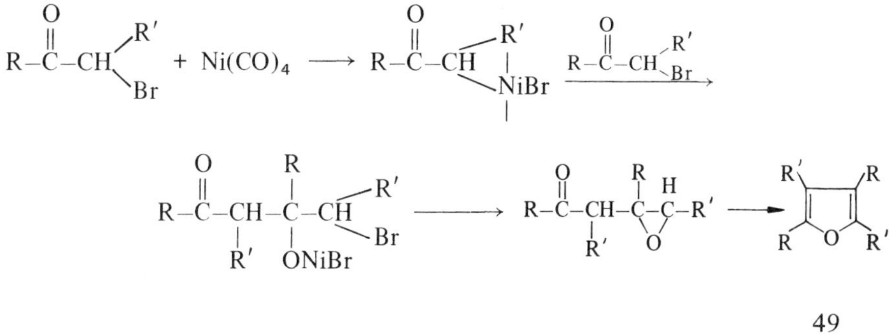

Similarly, Michael-type addition to the metal-carbon bond formed by the oxidative addition reaction is also possible [30])

$$\mathrm{Ph{-}CH_2Cl + CH_2{=}CH{-}CN \xrightarrow[Fe_3(CO)_{12}]{} Ph{-}CH_2{-}CH{=}CH{-}CN + Ph{-}CH_2{-}CH_2{-}CH_2{-}CN}$$

The possibility of displacement of an inert vinyl halogen with carbon monoxide was shown by Corey and Hegedus by the action of $Ni(CO)_4$ in the presence of a base [32]). The reaction can be explained by a similar mechanism.

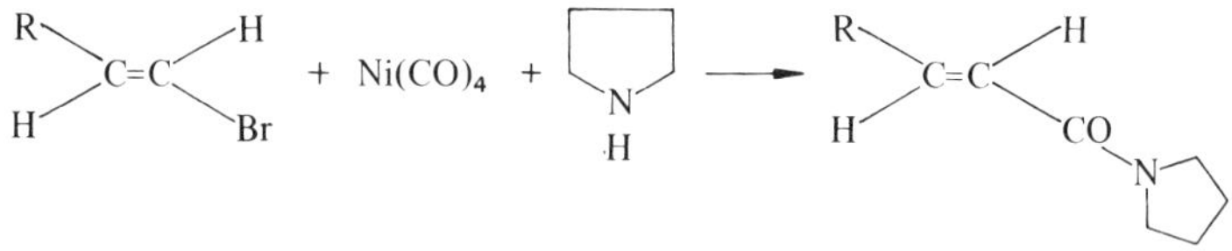

Certain metal carbonyls are very useful for the removal of halogens from *polyhalogenated compounds.* In these reactions, again, an oxidative addition step should be important. As shown below, carbenes [33]), cyclobutadiene [34]), and benzyne complexes [35]) are formed by the dehalogenation reaction with metal carbonyls.

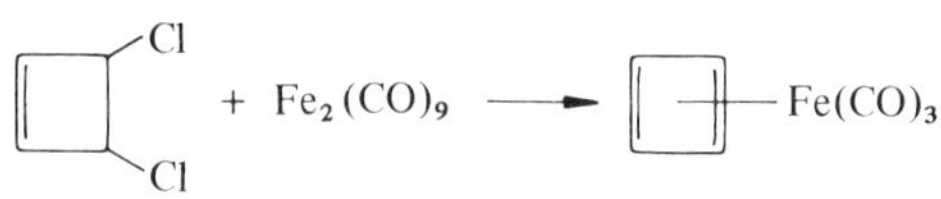

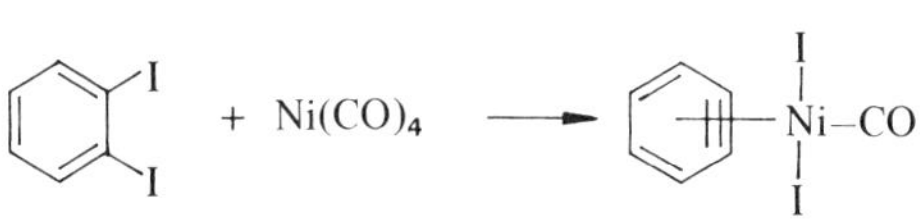

2. Reactivity of σ-Bonds

The above examples have shown how complexes having σ-bonds can be synthesized by oxidative addition reactions. An important factor to be considered in applying these oxidative reactions to organic synthesis is the reactivity of the σ-bond thus created. (Homolytic fragmentation, coupling and elimination reactions of σ-bonds are discussed in Chapt. IV.) Carbon on the electropositive metals (Mg, Li etc.) is usually anionic. In Grignard reagents, the reagents are assumed to have a carbanionic character. Their reactions can be explained by the nucleophilic character of the carbon directly bonded to magnesium. On the other hand, the characters of the transition metal-carbon and metal-hydrogen bonds formed by the oxidative addition (and insertion reactions described later) are much more versatile and cannot be predicted unequivocally. The character depends to a certain degree on the metallic species. For example, carbon bonded to nickel or iron in some complexes does behave as a nucleophile. However, whether carbon in a metal-carbon bond reacts with electrophiles or nucleophiles (or both) depends on many factors in addition to the identity of the metal. Factors such as oxidation state of metal, electronegativity and number of ligands, stability of reduced metal species, or medium, contribute to the characters of the metal-carbon bonds.

An example of the variation of the character of the carbon metal bonds is the *π-allyl complexes* of palladium and nickel. They are similar in structure. However, the nickel complex reacts with electrophiles such as aldehydes, ketones, and acrylonitrile [28,36,37].

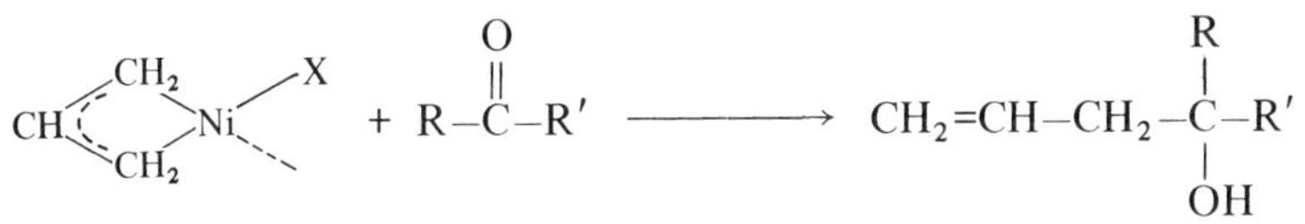

For example, two moles of acetone were linked to the terminal carbons of butadiene in the presence of a zero-valent nickel complex [38].

$$(COD)Ni + 6CH_3COCH_3 + CH_2{=}CH{-}CH{=}CH_2 \longrightarrow$$

$$2CH_3COCH{=}C(CH_3)_2 + 2COD + CH_3{-}C(CH_3)(OH){-}CH_2{-}CH{=}CH{-}CH_2{-}C(CH_3)(OH){-}CH_3$$

COD: cyclooctadiene

Furthermore π-allyl nickel complexes react with various alkyl, aryl, and vinyl halides in polar solvents such as dimethylformamide and N-methylpyrrolidone. Corey and Semmelhack studied the following reaction [37].

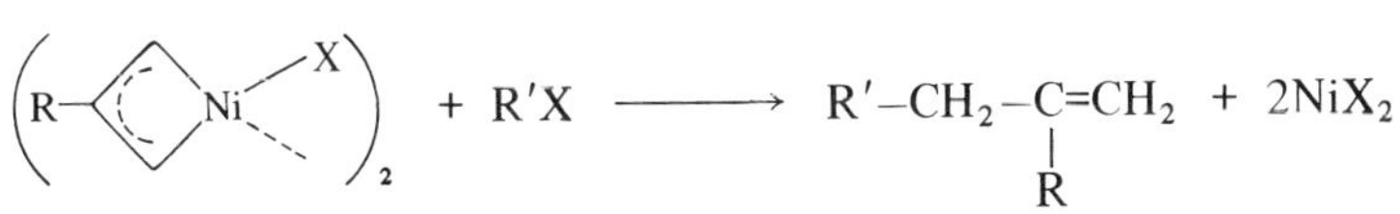

The synthesis of α-santalene from α,α-dimethylallyl bromide via π-allyl nickel complex is an elegant application.

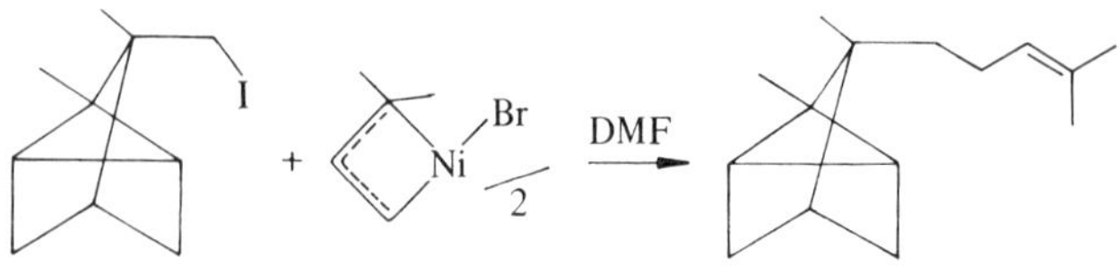

On the other hand, Tsuji, Morikawa, and Takahashi found that π-allyl palladium complexes react with nucleophiles, such as amines and enamines, and active methylene compounds, such as malonates, with reduction of bivalent palladium to the zero-valent metal [39,40]. This is due to the stability of the reduced species.

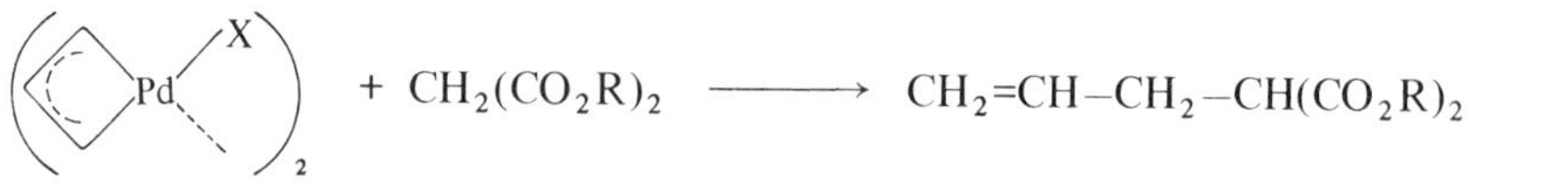

In the *dimerization reaction of butadiene* catalyzed by palladium complexes, nucleophiles (YH), such as amines, alcohols, phenols, carboxylic acids [41-45], and active methylene compounds [46] are introduced. This reaction can be explained by the attack of these nucleophiles on the π-allylic complexes formed as intermediates in the reactions. Takahashi, Shibano, and Hagihara confirmed by using deuterium that the hydrogen of YH migrates to C_6 of the dimeric product, probably via the oxidative addition reactions of YH to the palladium species [42].

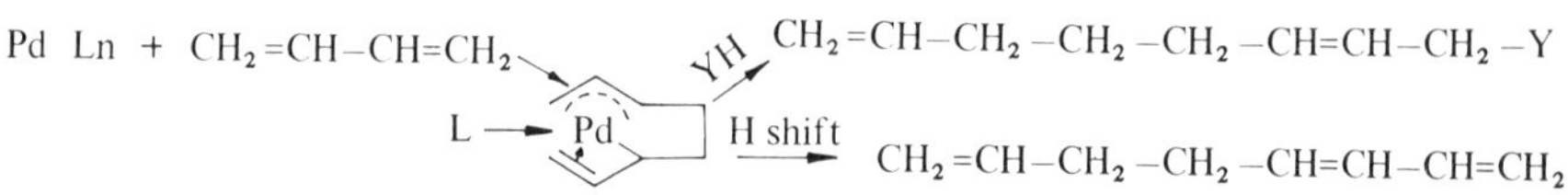

Butadiene does not always react with nucleophiles in the presence of palladium complexes. The reaction of aldehydes with butadiene in the presence of a catalytic amount of palladium-triphenylphosphine complex gave 1-substituted 2-vinyl-4,6-heptadien-1-ol, accompanied by 2-substituted 3,6-divinyltetrahydropyran [47,48]. In this reaction, apparently aldehydes behave as an electrophile.

As described before, nickel complexes react with *aldehydes* stoichiometrically. In view of the fact that aldehydes may undergo oxidative addition to palladium complexes, it is rather peculiar that ketones are not formed *via* oxidative addition. (This reaction will be discussed again in Chapt. V.)

Dimerization with incorporation of amines was observed also with nickel catalyst by Heimbach [49], but the reaction is less extensive. Thus, it can be said that the palladium complex is the most suitable for the addition of various nucleophiles to olefinic bonds. These examples show how different metals, such as palladium and nickel, are useful for different purposes.

Another important bond is the *metal-hydrogen bond.* Hydrogens bound to a transition metal have proton nmr absorptions in an unusually high range

(+5 to +18 ppm relative to tetramethylsilane), larger than in any other nontransition metal hydrides. This suggests that the hydrogen is highly shielded by the *d* electrons of the metal. The acidity and hence the reactivity of such hydrogen varies with the metals. For example, in various M-H bonds both H^+ and H^- characters are assumed from their reactions. The properties can be determined to a certain degree by the metallic species. However, even with the same metal, the properties of an M-H bond change with the other ligands attached to the metal. For example, $HCo(CO)_4$ is as strong an acid as mineral acids, but the electron density on the cobalt is increased with the coordination of phosphine instead of CO, and hence the acidity of the hydride becomes lower. Thus the dissociation constant of $HCo(CO)_3PPh_3$ is 1.09×10^{-7} and that of $HCo(CO)_3P(OPh)_3$ is 1.13×10^{-5} [50]. The acidity of cobalt carbonyl phosphine complexes decreases in the following order

$$HCo(CO)_4 > HCo(CO)_3(PPh_3) > HCo(CO)_2(PPh_3)_2 > HCo(CO)(PPh_3)_3 > HCo(dp)_2$$

dp: $Ph_2PCH_2{-}CH_2PPh_2$ [51].

It was found that the carbonyl stretching wave number of cobalt carbonyl phosphine complexes decreases with increase in the electron donating property of the phosphines [52].

The hydrogen of $HCo(dp)_2$ and $HCo(PBu_3)_4$ can be regarded as H^-, rather than H^+. Consequently, the reactivity or catalytic activity of these complexes differs. For example, the difference in *iso/normal* ratio in the oxo reaction was explained in terms of the difference in the ligands [53-56]. Thus, when a phosphine such as PBu_3 is coordinated to cobalt, the ratio of normal increases over that obtained with the corresponding carbonyl. In this case, a straight chain is formed by the olefin insertion when the acidity of the hydride is decreased. The increase of the normal aldehydes relative to *iso* aldehydes can be explained by the Markownikoff rule. Of course, as in many other cases, the steric effect, in addition to the electronic effect, should be considered at the same time.

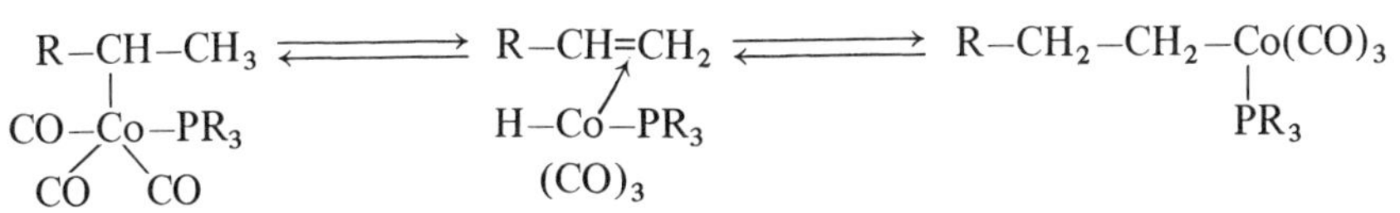

A characteristic and advantageous feature of the oxidative addition reactions of transition metal complexes compared with those of nontransition metal compounds is their *reversibility*. Grignard reactions cannot be made reversible, but the reverse reaction of oxidative addition is possible with transition metals in many cases. This reverse reaction is called *reductive elimination*. The ability of the transition metals to donate as well as to accept electrons easily from organic substrates is responsible for the reversibility. The first compound formed by oxidative addition is converted into secondary products by further reactions such as insertion reactions. The organic final product is then liberated by reductive elimination reactions with concomitant regeneration of the original complexes, thus making the whole process a catalytic cycle. If ligands are present which can stabilize an unstable intermediate complex, especially of low valence state, formed in the catalytic cycles, then the life and efficiency of the catalyst can be prolonged. (Reductive elimination is discussed in Chapt. IV.)

3. Oxidative Metal-Metal Bond Cleavage

Transition metal complexes can be classified into mononuclear and dinuclear (or polynuclear) complexes. Depending on the effective atomic number of the metal and the kind of ligands, the complexes can be mononuclear like $Fe(CO)_5$ or *dinuclear* like $Co_2(CO)_8$. The characteristic feature of dinuclear metal complexes is that they have a metal-metal bond in their structure. The metal-metal bond is cleaved oxidatively by the reaction with some covalent molecules to form σ-bonded complexes. This reaction is regarded as a one-electron oxidation. This is another way of forming σ-bonded complexes and is useful for organic synthesis.

$$-\overset{|}{\underset{|}{M}}-\overset{|}{\underset{|}{M}}- \;+\; A-B \longrightarrow -\overset{|}{\underset{|}{M}}-A \;+\; B-\overset{|}{\underset{|}{M}}-$$

The most typical example is the reaction of $Co_2(CO)_8$ with hydrogen in the first step of catalysis in the oxo reaction to form cobalt hydride. A similar reaction is the cobalt carbonyl-catalyzed hydrosilation reaction [57].

$$Co_2(CO)_8 + H_2 \longrightarrow 2HCo(CO)_4$$

$$Co_2(CO)_8 + HSiR_3 \longrightarrow HCo(CO)_4 + R_3SiCo(CO)_4$$

Similarly, several other covalent molecules, such as halogens, organometallic compounds, and carbon tetrachloride, take part in such reactions. Susuki and Tsuji reported that addition of carbon tetrachloride to olefins and carbonylation are catalyzed by dinuclear metal carbonyl complexes like $[\pi\text{-}C_5H_5Fe(CO)_2]_2$ and $[\pi\text{-}C_5H_5Mo(CO)_3]_2$ [58,59].

$$R-CH=CH_2 + CO + CCl_4 \longrightarrow R-\overset{\overset{COCl}{|}}{\underset{\underset{H}{|}}{C}}-CH_2CCl_3$$

In these reactions, the splitting of the metal-metal bond with carbon tetrachloride is assumed. The homolysis of the covalent carbon-chlorine bond is closely related to the formation of free radicals. Although the subject deviates somewhat from the topics of this chapter, there are some interesting synthetic reactions of *free radicals* generated by carbon tetrachloride and metal carbonyls. Dinuclear as well as mononuclear carbonyls are used for the generation of free radicals from *carbon tetrachloride*. For example, a radical type polymerization of various vinyl monomers is initiated with this system [60]. Radicals formed from the reaction of carbon tetrachloride and metal carbonyl are supposed to complex with the metallic species and are somewhat different in reactivity from radicals produced from the usual radical generators, such as peroxides. Mori and Tsuji found that the reaction of aniline and carbon tetrachloride in the presence of a catalytic amount of a metal carbonyl, such as $Co_2(CO)_8$ or $Mo(CO)_6$, gave *p*-amino-N,N′-diphenylbenzamidine in a high yield [61]. The formation of the amidine can be explained by the formation of a nitrogenous radical which migrates to a *para*-position of the aniline, followed by carbon tetrachloride attack.

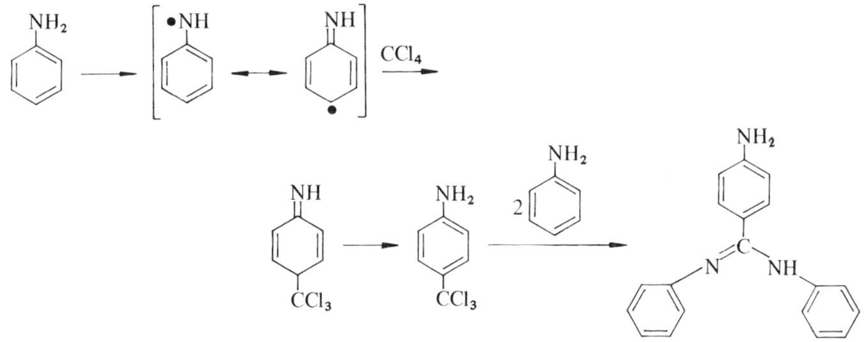

Another interesting reaction of carbon tetrachloride in the presence of metal carbonyl is the formation of triphenylimidazoles from benzylamine [62].

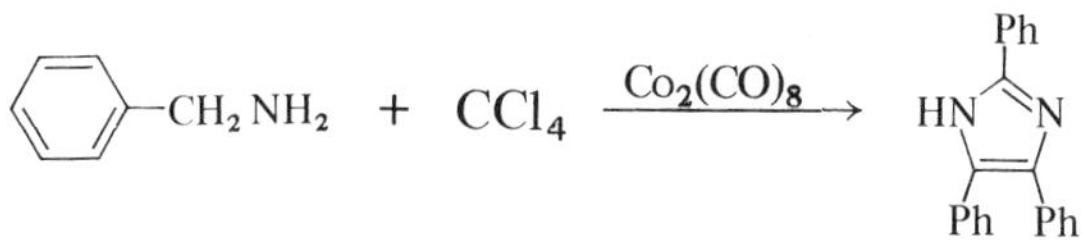

Oxidative cleavage of dinuclear metal carbonyls with tin compounds to form transition metal tin bond has been reported [63].

$K_4[Ni_2(CN)_6]$ has a dimeric structure with Ni-Ni bond [64]. The usefulness of this complex is apparent from its ready reaction with a variety of organic halides by splitting of the Ni-Ni bond. For example, the complex reacts with benzyl bromide to form a σ-bond and then undergoes coupling reaction or carbon monoxide insertion to give ketone [65,65a].

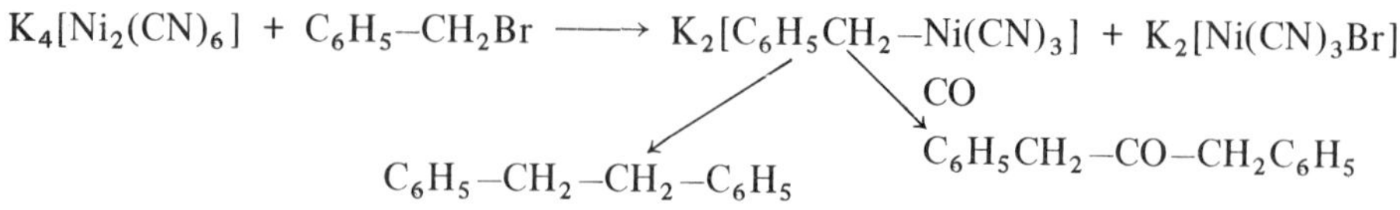

The direct cyanation of chemically inactive vinyl halides is also possible with the same complex under mild conditions [32].

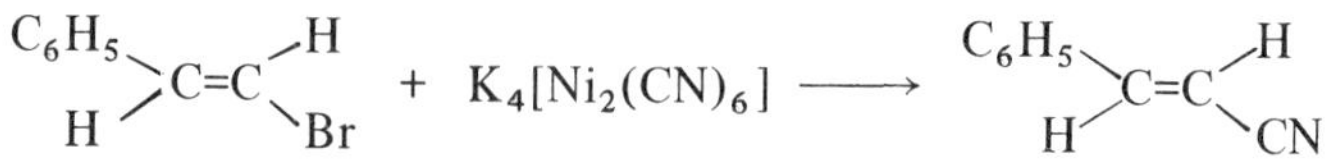

The reaction of *trans*-β-bromostyrene with the same complex in the presence of acrylonitrile in aqueous DMF gave 1,4-diphenyl-1,3-butadiene, β-cyanostyrene and 5-phenyl-4-pentenonitrile [66].

$$K_4Ni_2(CN)_6 + C_6H_5{-}CH{=}CH{-}Br + CH_2{=}CH{-}CN \longrightarrow$$

$$\underset{5.2\%}{C_6H_5{-}CH{=}CH{-}CH{=}CH{-}C_6H_5} + \underset{30.7\%}{C_6H_5{-}CH{=}CH{-}CN} + \underset{56.5\%}{C_6H_5{-}CH{=}CH{-}CH_2{-}CH_2{-}CN}$$

4. Reactions of Transition Metal Compounds with Organometallic Compounds

σ-Bonded alkyl-transition metal complexes can be prepared by the reaction of first-row transition metal compounds, such as halides, or acetylacetonates with organometallic alkylating agents, such as Grignard reagent, alkyl lithium, and alkyl aluminum, mercury, and zinc compounds [67].

$$L_nMX + R{-}M' \longrightarrow L_nM{-}R + M'X$$

The most famous example of this type of reaction is the *formation of the Ziegler catalyst,* useful for oligomerization and polymerization of olefins. In most cases, unstable alkyl or hydride complexes are formed and become very active catalysts. They are used without isolation. In the presence of certain ligands (carbon monoxide, CN, pyridine, bipyridine, triaryl, or trailkylphosphine), this reaction in some instances leads to stable isolable σ-bonded alkyl transition metal complexes. One example is the reaction of iron acetylacetonate with triethylaluminum in the presence of bipyridine to form a diethyliron complex [68-71].

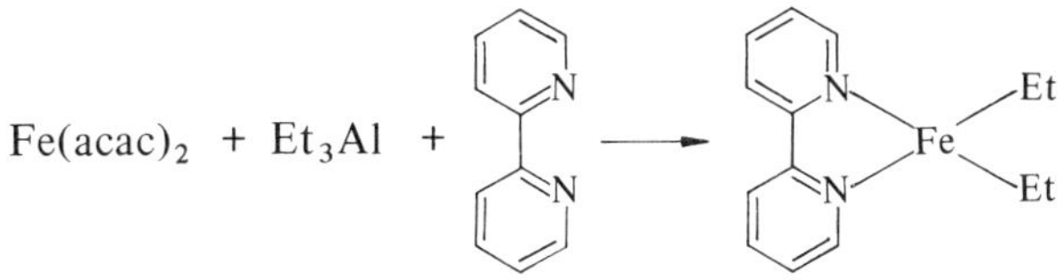

Wilkinson and coworkers obtained stable alkyl complexes by the reaction of transition metal halides or complex halides with trimethylsilylmethyl lithium [72]. With these alkyl complexes, no β-elimination reaction proceeds. The action of Grignard or organolithium reagents on halogen complexes of Pt(II) and Pd(II) has been widely used for the preparation of alkyl complexes, and quite stable alkyl complexes of noble metals can be prepared [73].

$$(Et_3P)_2PtMeI + MeMgI \longrightarrow (Et_3P)_2PtMe_2 + MgI_2$$

Organomercurials can also be used for the synthesis of alkyl complexes. Organomercurials have stable alkyl-mercury bonds, and the alkyl groups can be transferred to transition metal complexes (alkylation of the transition metal complexes). Thus diphenylmercury reacts with one equivalent of dichloroplatinum complex to produce phenylmercuric chloride and monosubstituted platinum complex in good yield [73].

$$(Me_2PhP)_2PtCl_2 + Ph_2Hg \longrightarrow (Me_2PhP)_2PtPhCl + PhHgCl$$

With carbonyl complexes, alkyl exchange often takes place with carbon monoxide insertion.

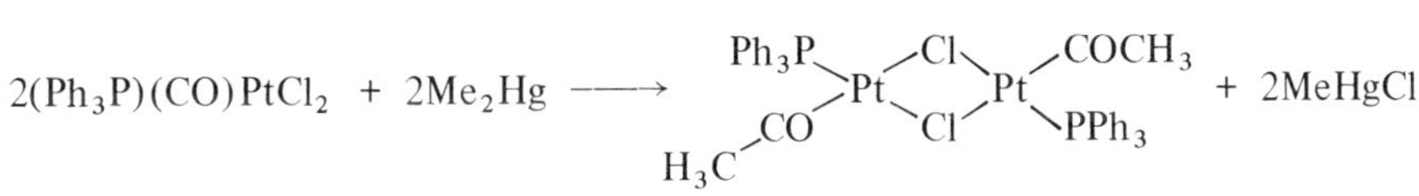

Wide application of this reaction to organic synthesis as a method of σ-bond formation has been found by Heck, using palladium salts [74-81]. Alkyl or aryl groups in mercurials are transferred to palladium to form alkyl or aryl palladium complexes, which are then used in situ for organic synthesis. Homogeneous solutions of palladium salts and alkyl or arylmercury compounds in polar solvents react to form solvated alkyl- or arylpalladium salts. These species are unstable and generally decompose at room temperature to coupled products, or to olefins if β-hydrogens are present in the alkyl groups. In the presence of olefins, however, alkyls without β-hydrogen, and aryls, add to the olefin, forming alkylethyl- or arylethylpalladium salts, which then rapidly decompose into palladium hydride and alkylated or arylated olefins [75].

$$PdX_2 + RHgX \longrightarrow RPdX + HgX_2$$

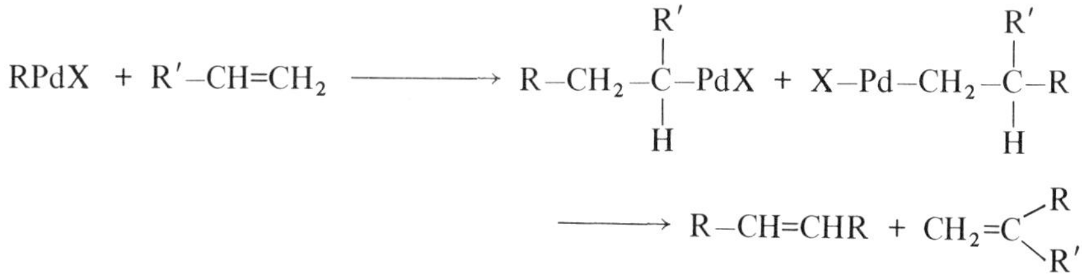

Thus, olefin alkylation and arylation are possible, and the scope of the reaction has been reported [81]. The reaction of propylene with phenyl-palladium acetate in methanol at 30 °C gave a 66% yield of a mixture of olefins containing 60% trans-propenylbenzene, 9% cis-propenylbenzene, 15% allylbenzene, and 16% α-methylstyrene [81].

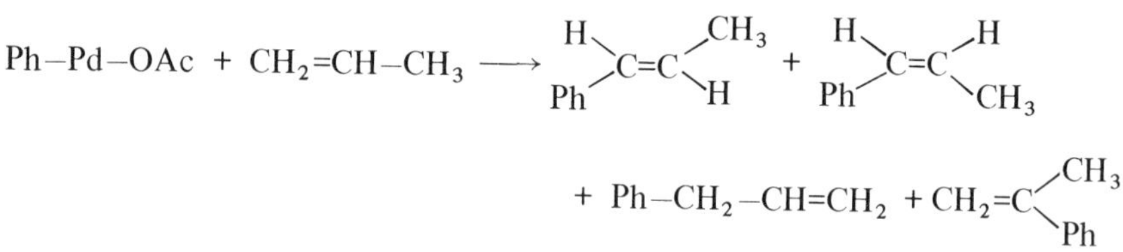

Arylpalladium salts, prepared in situ, react with allylic halides to give allylaromatic derivatives [76].

$$ArPdX + CH_2{=}CHCH_2X \longrightarrow ArCH_2CH{=}CH_2 + PdX_2$$

Reaction with primary and secondary allylic alcohols produces 3-arylaldehydes or 3-arylketones. This method provides a useful new route to a variety of 3-arylaldehydes or -ketones [75].

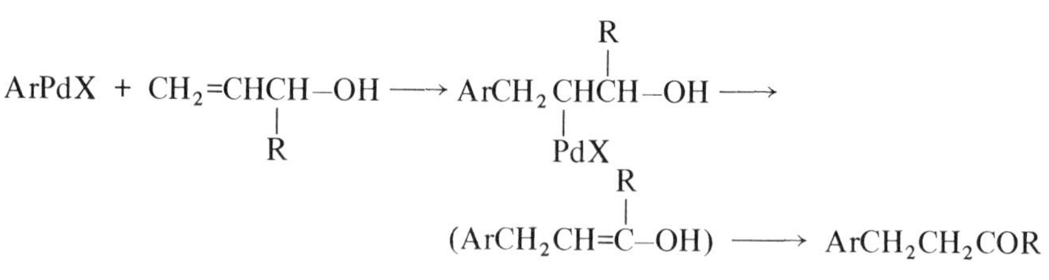

Reaction of carbomethoxymercuric acetate with palladium acetate gives „*carbomethoxypalladium salt*", which can be used for the carbomethoxylation of olefins to form unsaturated esters. For example, the reaction with α-methylstyrene produced 86% yield of unsaturated esters, 96% of this being the non-conjugated methyl 3-phenyl-3-butenoate and only 4% the *trans*-conjugated product [81].

$$(CH_3{-}OCO)Hg(OAc) + Pd(OAc)_2 \longrightarrow (AcO)Pd(OCO{-}CH_3)$$

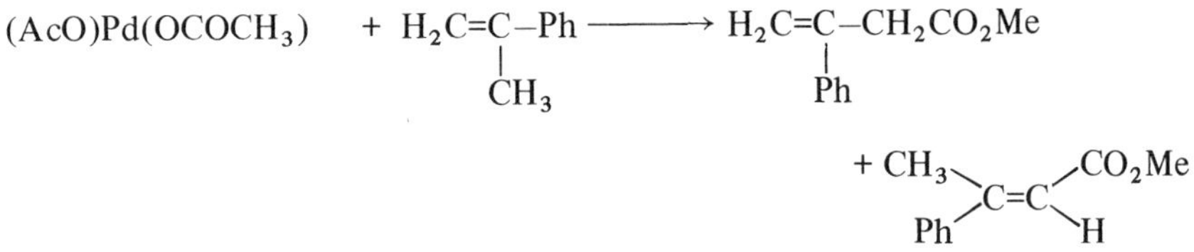

Henry found that arylpalladium salts react with carbon monoxide in hydroxylic solvents to form arylcarboxylic acids or their derivatives, depending upon the solvents [82].

$$C_6H_5PdX + CO \longrightarrow C_6H_5COX$$

Under some conditions, diaryl ketones are formed in moderate yield from arylmercuric salts and carbon monoxide with a palladium salt catalyst [80].

Seyferth and Spohn found that organomercury compounds and $Co_2(CO)_8$ react in THF to give ketones in moderate yields [83,84].

$$RHgX \text{ or } RHgR + Co_2(CO)_8 \longrightarrow RCOR + Hg[Co(CO)_4]_2$$

Although the mechanism of the reaction is not completely clear, it is certain that, organomercurials behave as alkylating agents of the cobalt complex. In improving this reaction, Seyferth found that carbonylation of mercurials can be carried out in the presence of catalytic amounts of $Co_2(CO)_8$ or $Hg[Co(CO)_4]_2$ under UV irradiation.

$$R_2Hg + CO \xrightarrow[Co_2(CO)_8]{UV} RCOR + Hg$$

5. Formation of Metal Carbonylate Anions from Metal Carbonyls and Their Reactions

Reactive organic halogen compounds add oxidatively to metal carbonyls affording useful intermediates for organic synthesis, as discussed before. The reactivity of metal carbonyls can be further enhanced by converting them into metal carbonylate anions with certain alkyl- or arylmetal compounds, such as alkyl- or aryl compounds of lithium, mercury, and tin. The reactions of carbonyls of nickel, iron, molybdenum, tungsten and chromium have been studied extensively [85,86]. Organolithium compounds are very reactive toward metal carbonyls and add reductively to the metal, even at low temperature, to form rather stable anionic complexes. The products of the reaction with $Ni(CO)_4$ and $Fe(CO)_5$ are most useful for organic synthesis. They are called lithium aroyl- or acylmetal carbonylates. Extensive studies in this area have been carried out by Ryang, Tsutsumi, and their coworkers [87]

$$\mathrm{RLi} + \mathrm{M(CO)}_n \longrightarrow \mathrm{Li[RCO{-}M(CO)}_{n-1}]$$

M: Ni, n = 4 or Fe, n = 5

The aroyl carbonylates readily decompose, giving various carbonyl compounds. The most useful property of the aroyl carbonylates is their nucleophilic character. In the usual organic chemistry, the acyl group is electrophilic and no example of a nucleophilic reaction of the acyl group is known [88]. On the other hand, the acyl group attached to the metal in aroyl carbonylates is nucleophilic, thus unique reactions of nucleophilic groups are possible. Some typical examples are shown below.

Hydrolysis of the aroylironcarbonylates gives aldehydes [89]. This method is especially useful for the synthesis of aldehydes which contain unsaturated groups [90].

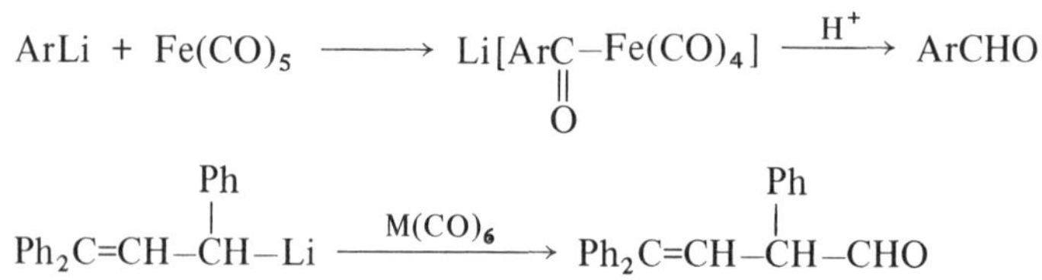

Unlike the corresponding iron complexes, aroylnickel carbonylates give acyloins instead of aldehydes by hydrolysis with aqueous methanol containing hydrochloric acid. In reactions at higher temperature, or by treatment with bromine, α-diketones are formed [85].

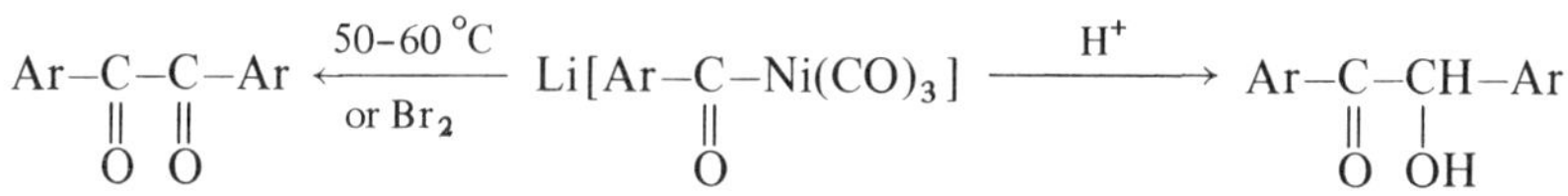

The nucleophilic character of the complexes is apparent from the reactions with alkyl halides or acyl halides. With the iron complex, unsymmetrical ketones are formed [91].

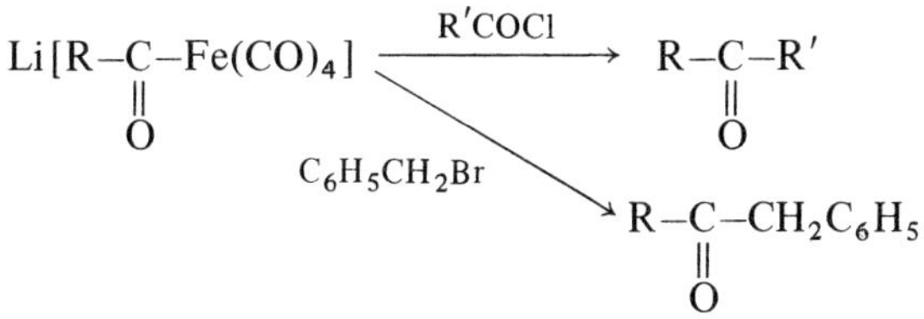

On the other hand, nickel complexes give acyloins or en-diol diesters.

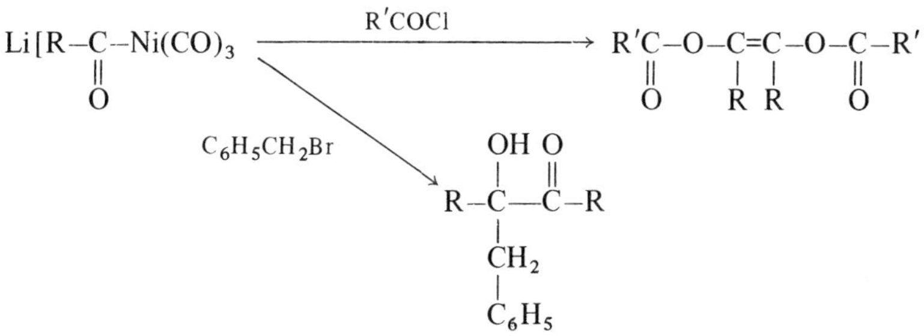

It should be noticed that iron complexes produce mostly monocarbonyl compounds, while nickel complexes give dicarbonyl compounds. The difference between nickel and iron was explained by the structures of the mononuclear iron complex and the dinuclear nickel complex formed by the reactions of the metal carbonyls with phenyl lithium as shown below [92].

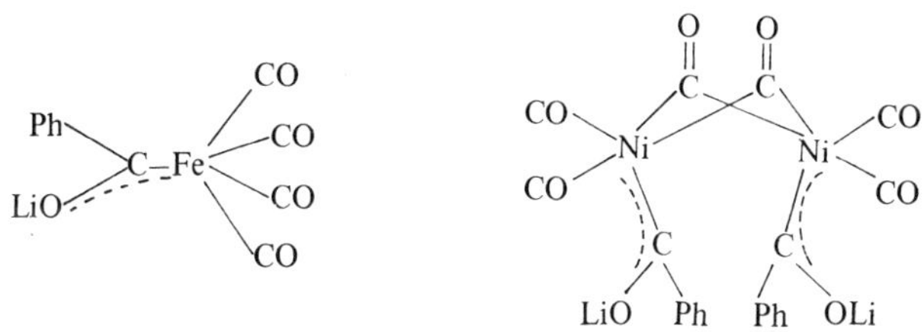

As expected from their nucleophilic character, also Michael addition-type reactions are possible to form 1,4-dicarbonyl compounds [93].

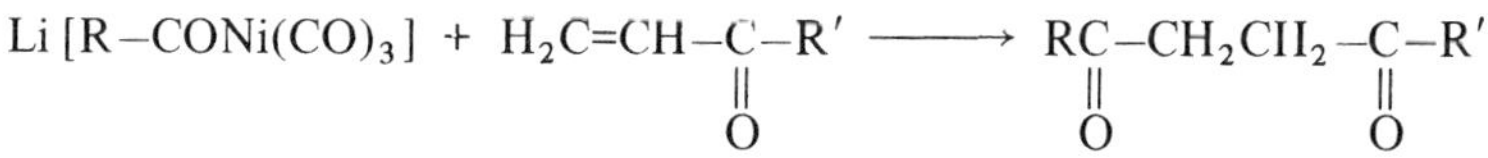

1,4-Dicarbonyl compounds are also formed by the reaction of acetylenic compounds with nickel carbonylates [94].

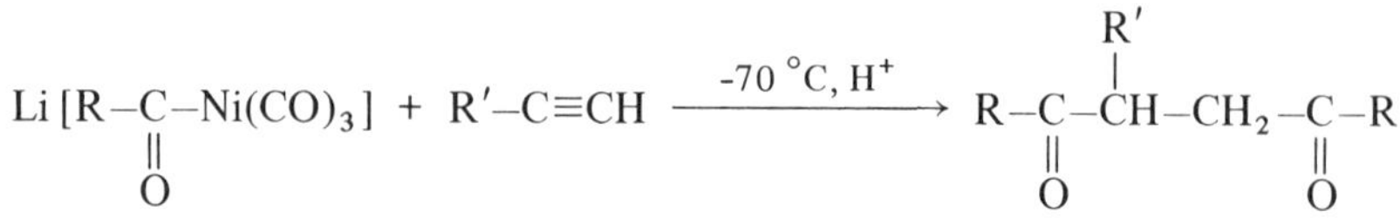

Lithium carbamoylnickel carbonylate is formed from the reaction of lithium amide and $Ni(CO)_4$. This complex gives with phenylacetylene 2-phenyl-N,N.N′,N′-teramethylsuccinamide as a main product and N,N-dimethylcinnamamide as a minor product [95].

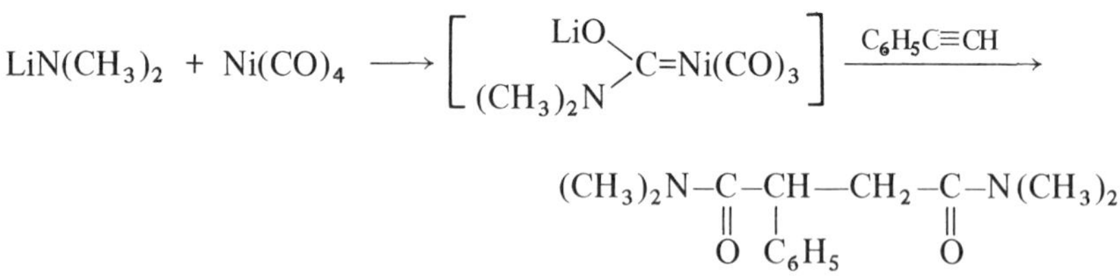

When the complex is treated with organic halides or acyl halides, dimethylaminocarbonylation proceeds smoothly. For example, reaction with benzyl bromide gave N,N-dimethylphenylacetamide in 64.5 % [96].

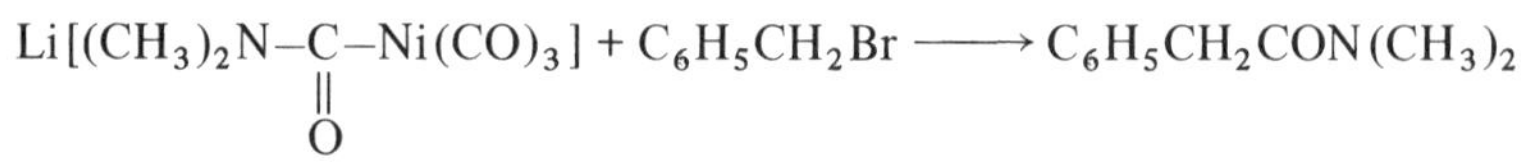

Another way of forming carbonylate anion is the reaction of $Ni(CO)_4$ with KOR, especially potassium *t*-butoxide. The carbonylate complex thus formed undergoes butoxycarbonylation of moderately active organic halides to give a *t*-butyl ester [32].

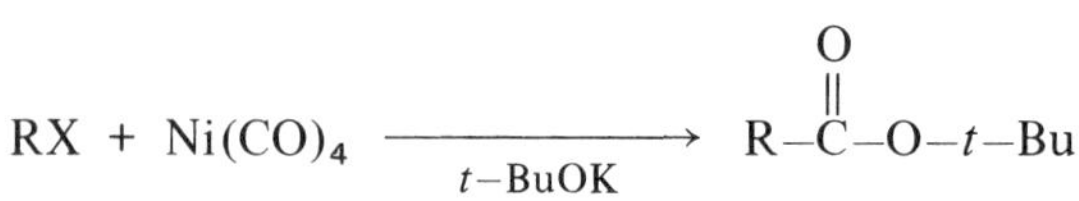

Another example is the conversion of alkyl bromides into aldehydes using $Fe(CO)_5$ [97]. $Fe(CO)_5$ is reduced by sodium amalgam and the result is a coordinatively unsaturated iron dianion. Oxidative addition of alkyl bromide affords the saturated alkyl iron anion. Alkyl-acyl rearrangement takes place by the addition of triphenylphosphine to give acyl iron anion. Protonation of the anion results in the intermediate acyl iron hydride which undergoes reductive elimination to yield aldehydes.

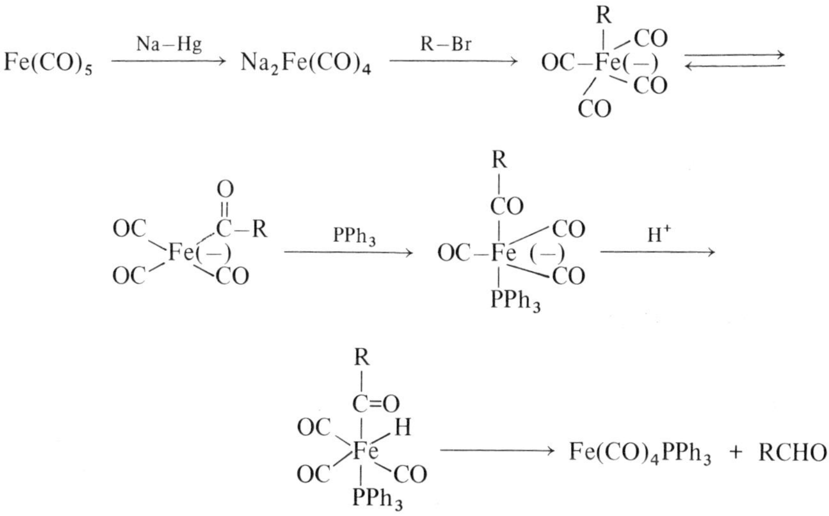

The reactivity of carbonyls of Cr, Mo, and W with alkyl lithium is somewhat lower than that of nickel and iron carbonyls. The most interesting reaction of these metal carbonyls, found and studied extensively by Fischer and his coworkers, is the formation of *carbene complexes* [98,99]. Phenyl or methyl lithium adds readily to $W(CO)_6$ at room temperature to form carbonylate complexes. The tetramethyl ammonium salt of the complex is treated with acid and then with diazomethane to give a neutral complex whose structure was found to be that of a carbene complex. In this way a series of carbene complexes of Cr, Mo, and Mn were synthesized.

$$RLi + W(CO)_6 \longrightarrow Li[W(CO)_5COR]$$

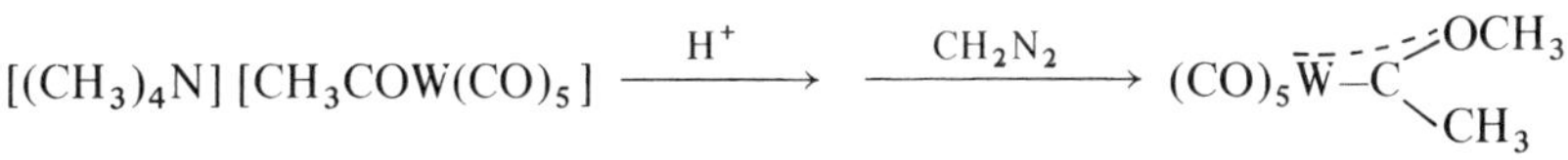

The reaction of these transition metal carbene complexes with some nucleophiles such as isocyanante, thiophenol, hydrazine, or hydroxylamine, have been studied. For example, the carbene part of the complex is converted into vinyl ether by pyridine [100].

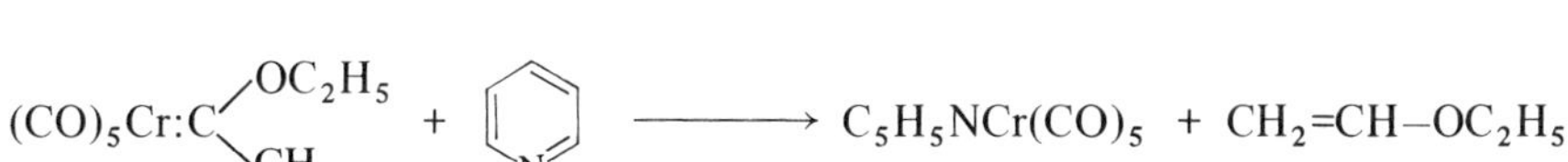

The reaction of the carbene complex with an activated olefin produced a cyclopropane derivative [101].

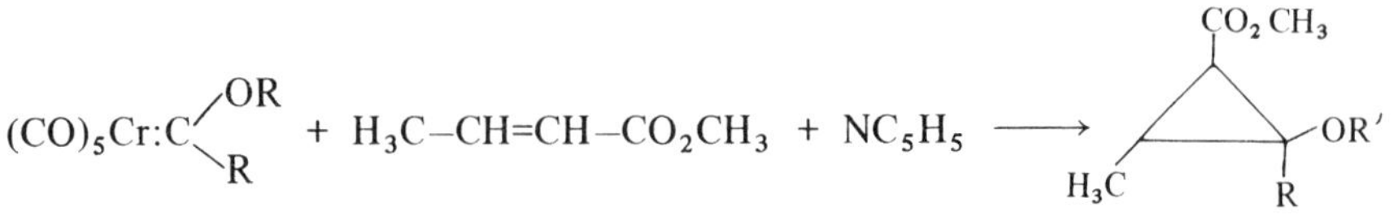

III. Insertion Reactions

The reactions described in the preceding chapters enable various metal complexes having σ-bonds to be formed. In order to form more elaborate organic compounds from these complexes, further transformations may be achieved by the addition of other molecules. This type of addition reaction is called "insertion". There are many types of insertion reactions, depending on the σ-bonds and the molecules to be inserted in them. Examples of the participants in these reactions are shown in Table 1.

Table 1

σ-Bond	Bonds and molecules to be inserted	
M–H	C=C	N=N
M–C	C=C–C=C	C=N
M–X	C≡C	
M–M′	C=O	
M–N	CO, CO_2, SO_2	Carbene
M–O	RNC	

These combinations suggest a large number of possible insertion reactions, but not all of them are in fact possible.

The insertion of *olefins* is the most common, and insertions in M-H or metal alkyl bonds actually take place in the homogeneous hydrogenation, oligomerization, polymerization of olefins, etc.

Many reactions of Pd(II) salts, useful for organic synthesis, have been described [103-107]. Most of those involving olefins can be explained by the insertion reaction. The most remarkable property of Pd(II) is the substitution of vinyl hydrogen or halogens by nucleophiles. The usefulness of this reaction is apparent from the consideration that addition to olefins is common in usual organic reactions but no displacement reaction of olefinic hydrogen is possible. By coordination with palladium salts, nucleophilic displacement on olefins becomes feasible. When coordinated with Pd(II), an olefinic bond reacts with nucleophiles and olefin insertion takes place. The olefin insertion reaction in this case is called a *"palladation reaction"*. In palladation reactions, a ligand A on palladium migrates to π-bonded olefin which simultaneously is transformed into a σ-bonded complex. Thus, by the reaction of palladium olefin complexes with nucleophiles which contribute –OH, –OR, –OCOR, $-NR_2$, $-N_3$, carbanions, and others, it is possible to prepare vinyl compounds by substitution on vinyl carbons. In some cases, a nucleophilic addition which leads to bifunctional products is possible (nucleophilic addition). There are numerous examples which can be explained by this mechanism.

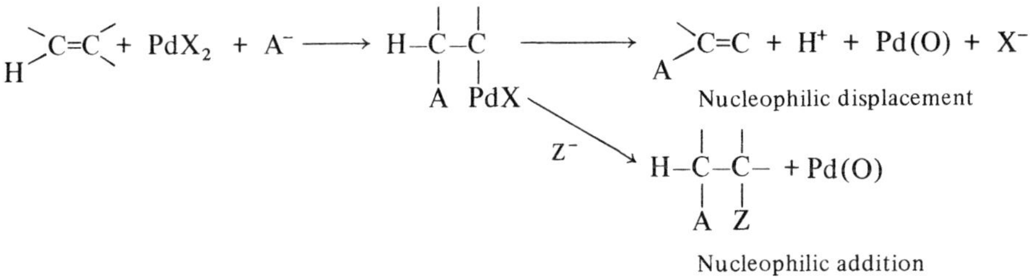

Well known is the reaction of palladium which leads to the formation of carbonyl compounds, vinyl acetate and vinyl ether by the reaction of olefins with water, acetate, and alcoholate.

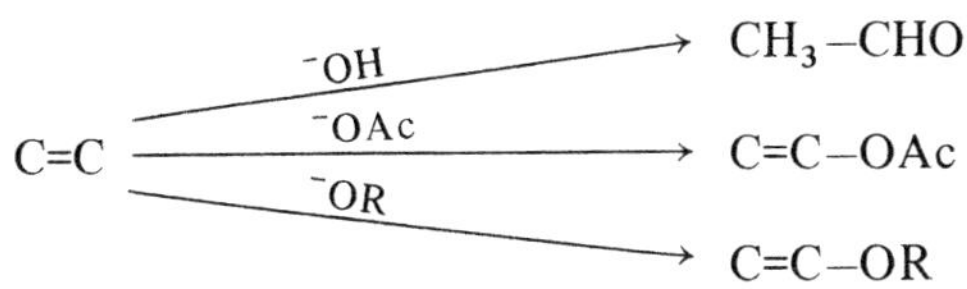

The product of the palladation reaction exists as an active intermediate and cannot be isolated in general. However, the product of palladation was isolated as a stable compound in the reaction of cyclooctadiene palladium complex with carbanions such as malonate or alcoholate. Further reaction of the complex with base to give bicyclo (6,1,0) nonene and bicyclo (3,3,0) octane systems was reported by Takahashi and Tsuji [108]. The reactions are understood as intra- and intermolecular nucleophilic addition reactions.

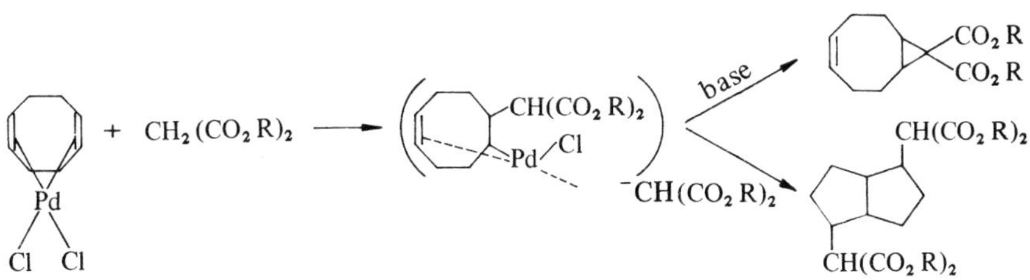

σ-Bonds between aromatic rings and transition metals can be synthesized by several means [109]. These bonds are susceptible to insertions. Typical examples are observed with palladium complexes. The reaction of arylmercurials and palladium salts followed by olefin insertion, as described before, is a typical example [74]. Direct aromatic substitution by reactions with olefins, discovered by Moritani and Fujiwara, in the presence of palladium salts also proceeds through olefin insertion [110].

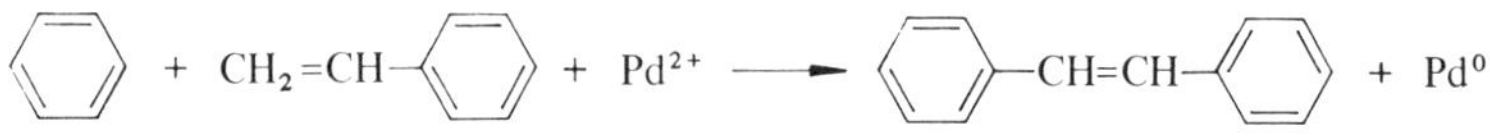

The following example is evidence of facile olefin insertion into a benzene-palladium σ-bond [103].

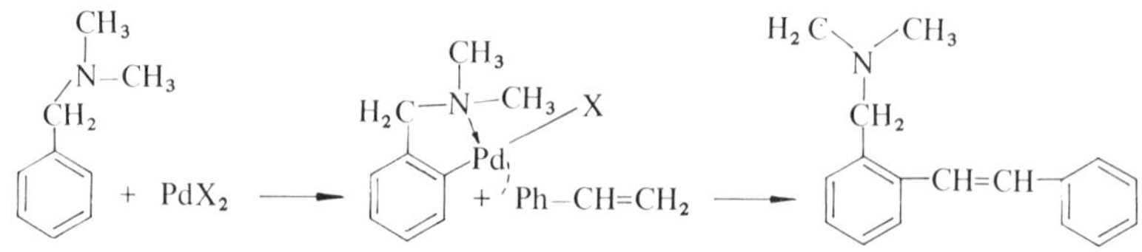

Carbon monoxide insertion is another widely observed reaction. The following carbon monoxide insertion into an azobenzene palladium complex is typical [111]. Azobenzene-palladium chloride complex has a carbon-palladium bond at the *ortho*-position, the carbonylation of which gives indazolinones by carbon monoxide insertion. Further carbonylation of indazolinone catalyzed by $Co_2(CO)_8$ produces quinazolinedione, hydolysis of which gives aniline and anthranilic acid [172]. Thus this method is useful for the synthesis of substituted anthranilic acid.

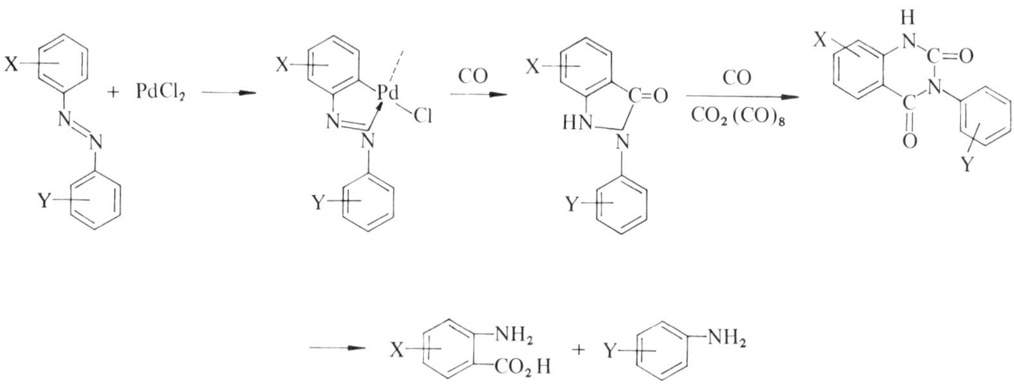

The reaction of chlorine with the azobenzene complex affords *ortho*-chlorinated azobenzene, and palladium chloride is regenerated [112].

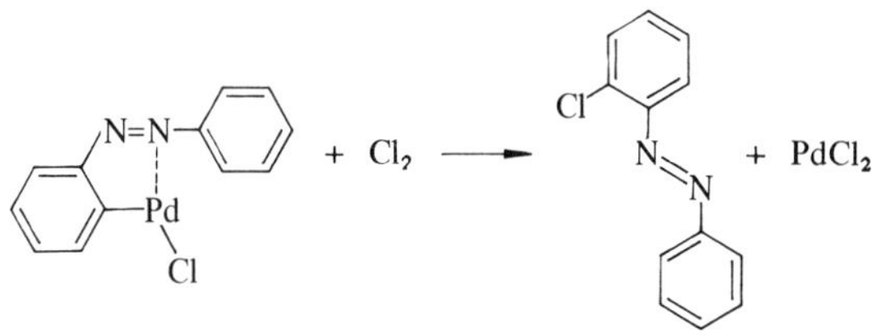

These reactions are therefore useful for the *ortho*-substitution of aromatic rings.

Isocyanide is isoelectronic with carbon monoxide, and an insertion of isocyanide, similar to carbon monoxide, is well known [113,114]. Thus π-allylpalladium chloride reacts with cyclohexyl isocyanide. Alcoholysis of the insertion product gives N-cyclohexyl-3-butenimidate [115].

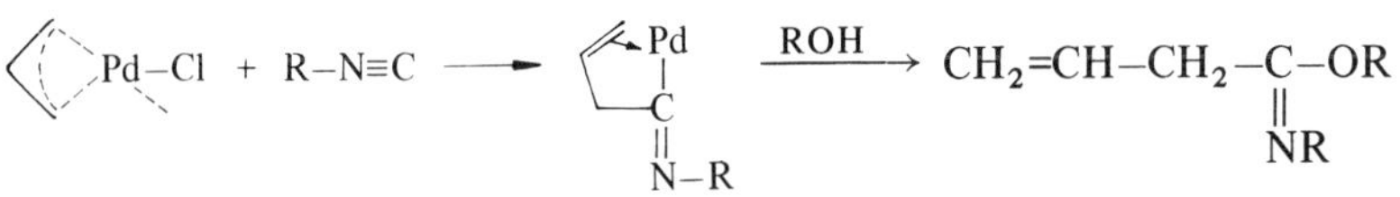

Acylmetal complexes generally react with conjugated dienes to produce useful organic compounds. For example, acylcobalt carbonyls add to butadiene, producing 1-acylmethyl-π-allylcobalt derivatives, which then undergo elimination of the elements of $HCo(CO)_3$ on treatment with base to give 1-acyldiene [116]. This reaction can be made catalytic with respect to the cobalt catalyst under proper reaction conditions.

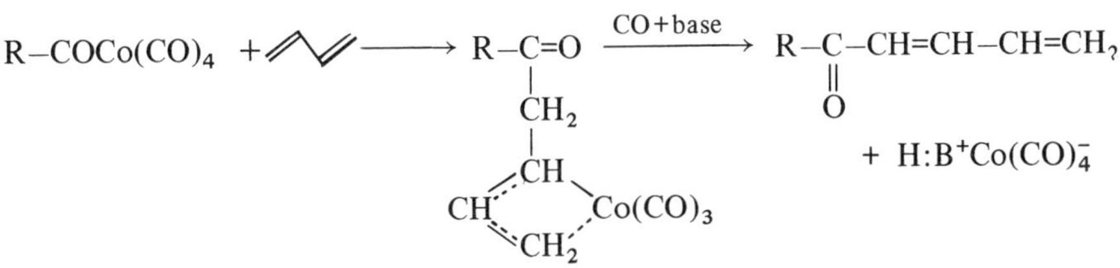

Klein reported the formation of *trans*-2-butenylsulfone by the reaction of ethylene and sulfur dioxide in the presence of palladium salts as catalysts [117]. The reaction starts with the insertion of ethylene into a Pd-H bond, followed by sulfur dioxide insertion [118]. Then two molecules of ethylene are inserted into the palladium sulfur bond.

$$CH_2{=}CH_2 + H{-}Pd{-}X \longrightarrow CH_3{-}CH_2{-}Pd{-}X \xrightarrow{SO_2} CH_3{-}CH_2{-}SO_2{-}Pd{-}X$$

$$\xrightarrow{2CH_2=CH_2} CH_3{-}CH_2SO_2CH_2{-}CH_2{-}CH_2{-}CH_2{-}Pd{-}X \longrightarrow$$

$$CH_3{-}CH_2SO_2CH_2{-}CH_2{-}CH{=}CH_2 \longrightarrow CH_3{-}CH_2SO_2CH_2{-}CH{=}CH{-}CH_3$$

Carbon dioxide is abundant and readily available, but its reaction with transition metal complexes has not been extensively studied. A few examples of carbon dioxide insertion are known. Thus, formic acid can be formed by the insertion of carbon dioxide into the cobalt hydride bond [119,120].

$$H(N_2)Co(PPh_3)_3 + CO_2 \longrightarrow H{-}\overset{\overset{\displaystyle O}{\|}}{C}{-}OCo(PPh_3)_3 + N_2 \xrightarrow{CH_3I} H{-}CO_2CH_3$$

In the formation of formamide by the reaction of amine, hydrogen, and carbon dioxide using Rh, Ir, and Co catalysts, the insertion of carbon dioxide is assumed [121].

$$L_xMH + CO_2 \longrightarrow L_xM{-}\overset{\overset{\displaystyle O}{\|}}{C}{-}OH \xrightarrow{HN(CH_3)_2} L_xMOH + H{-}\overset{\overset{\displaystyle O}{\|}}{C}{-}N(CH_3)_2$$

$$L_xMOH + H_2 \longrightarrow L_xMH + H_2O$$

So far no definite carbon dioxide insertion into analogous metal carbon bonds has been reported. A rhodium complex coordinated with carbon dioxide was reported by Iwashita and Hayata [122].

Carbene is an active species, the stabilization of which by coordination was described earlier. The coordinated carbene also undergoes insertion reactions. For example, bis(trifluoromethyl)carbene from the diazo compound was inserted into a Mn-H bond to give an alkyl complex [123].

$$(CF_3)_2CN_2 + HMn(CO)_5 \longrightarrow (CF_3)_2CH{-}Mn(CO)_5$$

In the reaction shown below, the carbene generated from diazoacetic acid was inserted into the Ni-C bond to give pentadienoate [124].

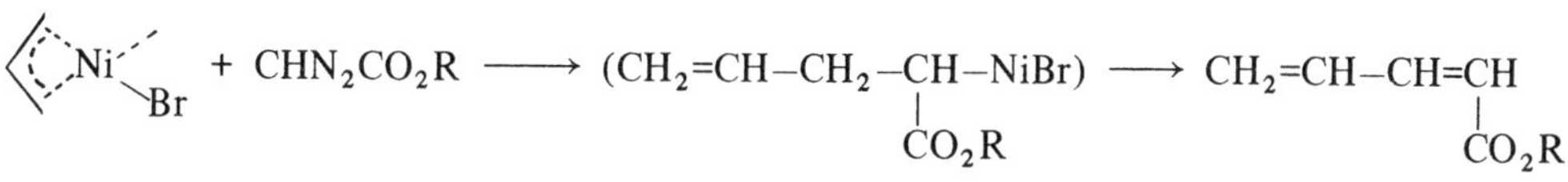

The insertion reactions may occur more than once. *Successive insertions* can make quite complex organic compounds. For example, successive olefin insertion takes place in the polymerization reaction. The factors which differentiate oligomerization and polymerization are not clear. That is, the factors which determine whether one step or successive insertions will occur constitute an interesting but difficult problem. Selective and successive insertion reactions were observed by Chiusoli, Pottaccio, and Cassar in the reaction of allyl chloride, acetylene, and carbon monoxide in the presence of $Ni(CO)_4$ [125,126]. A π-allyl-nickel complex is formed first by the reaction of allyl chloride and $Ni(CO)_4$, then acetylene insertion takes place. Then follows carbon monoxide insertion to form a cis-2,5-hexadienoyl complex, which cyclizes to a five membered ketone by the insertion of terminal olefin into the Ni-carbon bond. Again carbon monoxide insertion takes place to form an acyl complex. If the reaction is terminated at this stage with water, the product is 2-oxocyclopent-3-enyl-acetic acid.

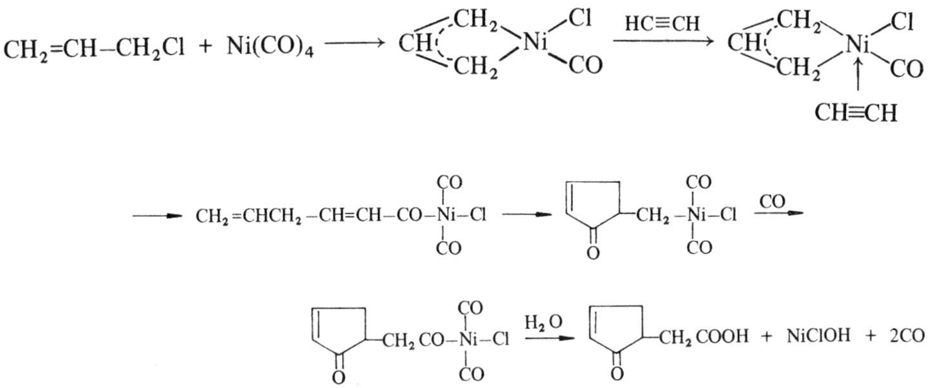

In acetone, further reactions with acetylene and carbon monoxide continue to form another unsaturated acyl complex. The last step is the formation of an unsaturated lactone. This reaction is understood to be an intramolecular insertion of the ketonic carbonyl into the Ni-carbon bond.

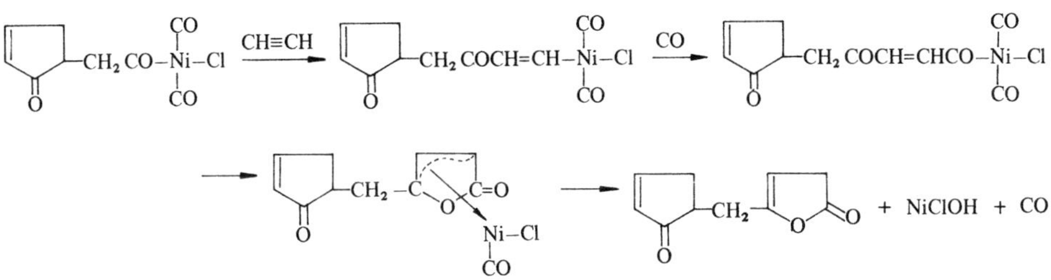

In addition, acetone itself reacts with the intermediate nickel complex to form the following compounds. This reaction is similar to the Reformatsky reaction.

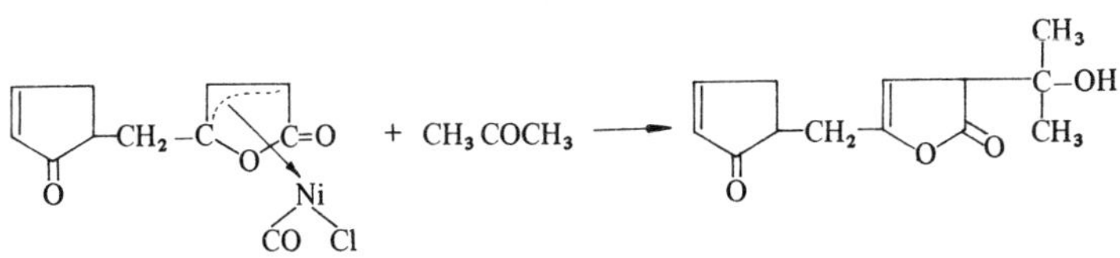

This is a remarkable example of the way in which complex molecules can be fabricated from simple molecules by a series of orderly insertion reactions.

IV. Liberation of Organic Compounds from the σ-Complexes

The last step in the synthetic organic reactions via transition metal complexes is the liberation of the fabricated organic compounds from the complexes. Coordinative lability of the ligands is responsible for this step. The general pattern of reactions of the σ-alkyl bond to liberate organic products can be summarized in the following way.

$-M-CH_2CH_2R$ →
- → CH_3CH_2R (hydrogen abstraction)
- → $RCH_2CH_2CH_2CH_2R$ (coupling)
- → $CH_2{=}CHR + CH_3CH_2R$ (disproportionation)
- → M–L (low valent state)
- → $CH_2{=}CHR$ + M–H (metal hydride formation)

The initially formed unstable alkyl transition metal compounds mostly undergo homolytic fragmentation and elimination reactions leading to alkane, alkene, dimeric hydrocarbons plus metal hydride and colloidal or active metal [168,169]. The stability of the alkyl transition metal compounds increases, for example, in the order Ni ⟨ Pd ⟨ Pt.

The most common reaction is coupling of the coordinated ligands. Usually an entering ligand or accelerating ligand labilizes the complex and the coupling takes place between groups already bonded to the metal. A typical example of this type is the *formation of cyclododecatriene* by the accelerating action of phosphine on the intermediate open chain complex [127].

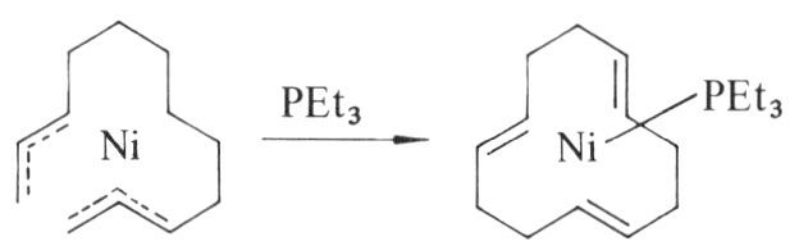

In the catalytic formation of cyclododecatriene, the open chain species is cyclized and displaced by entering butadiene with simultaneous reformation of the open chain species. When carbon monoxide is added to the complex, coupling takes place with incorporation of carbon monoxide to form cyclic ketone, namely 11-vinyl-3,7-cycloundecadien-1-one [128]. Isocyanide is isoelectronic with carbon monoxide, as described before; thus isocyanide effects similar intramolecular cyclization of the complex. Hydrolysis of the cyclized products gave 3,7,11-cyclotridecatrien-1-one and 11-vinyl-3,7-cycloundecadien-1-one.

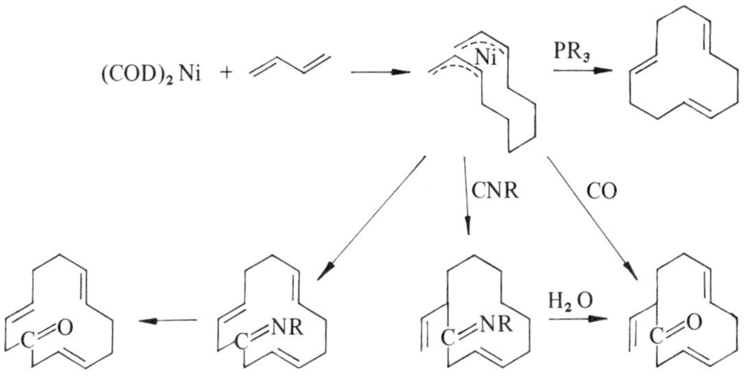

The cyclization of butadiene, as illustrated, involves the coupling of the π-allylnickel moiety. Various cyclic 1,5-dienes can also be synthesized by the coupling reaction of allyl dibromides using $Ni(CO)_4$ via a π-allyl complex. Elegant applications of this method to the syntheses of large ring (12, 14, and 18) compounds and natural products have been reported by Corey and Wat [129]. The last step of the humulene synthesis by Corey and Hamanaka is the following coupling reaction of allylic dibromide via π-allylnickel complex formation [130].

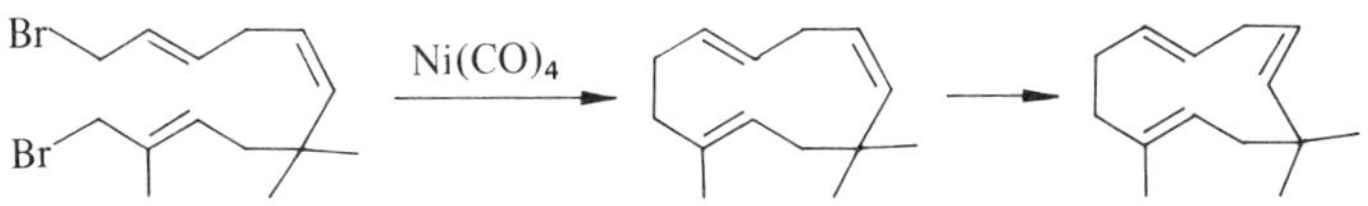

π-Allylpalladium acetylacetonate undergoes the coupling reaction of the ligands due to the labilizing action of the entering carbon monoxide to form allylacetylacetone [131].

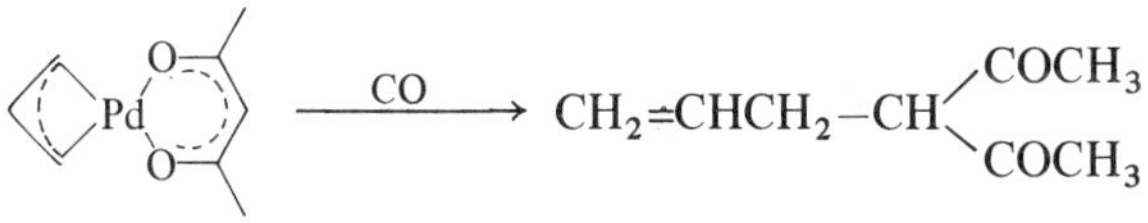

Mechanistic and kinetic studies on the activation of metal-alkyl bonds by entering olefins are carried out by Yamamoto, Yamamoto, and Ikeda [132]. For example, diethyldipyridylnickel is a quite stable complex up to 100 °C. However, thc formation of butane by ligand coupling, even at low temperature, was observed when the complex was treated with olefinic compounds such as acrylonitrile. Blue shift in a spectrum was observed when acrylonitrile was added to the diethyl complex. This change was explained by back donation from the nickel to the olefins, and it was concluded that the stronger the coordination of the entering olefins, the more the alkyl-metal is activated.

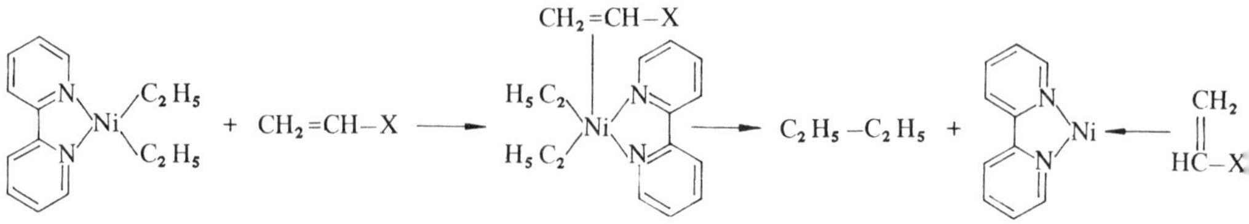

The acrylonitrile coordinated diethyldipryridyl nickel complex was isolated as an unstable intermediate [133].

These labilizations of the ligands to cause coupling may occur due to the overfilling of the coordination sphere. There are many other examples of coupling reactions.

Another type of final step is the formation of *hydride-olefin complexes* by a β-elimination reaction. Metal-alkyl complexes undergo this reaction. The driving force of the reaction may be initial coordinative unsaturation. In the catalytic cycle, olefin in the complex formed by the β-elimination is displaced by the entering olefins. The elimination is the reverse process of the olefin insertion reaction, and is called a σ-π-rearrangement [134]. The termination step in the oligomerization and polymerization of olefins is an example of this reaction. The regenerated hydride adds fresh olefin to act as catalyst again. The Scheme 1 shown for addition of ethylene to butadiene illustrates the sequence [135]. In this reaction, hydride formed by β-elimination is transferred to coordinated butadiene to give a π-allylic complex.

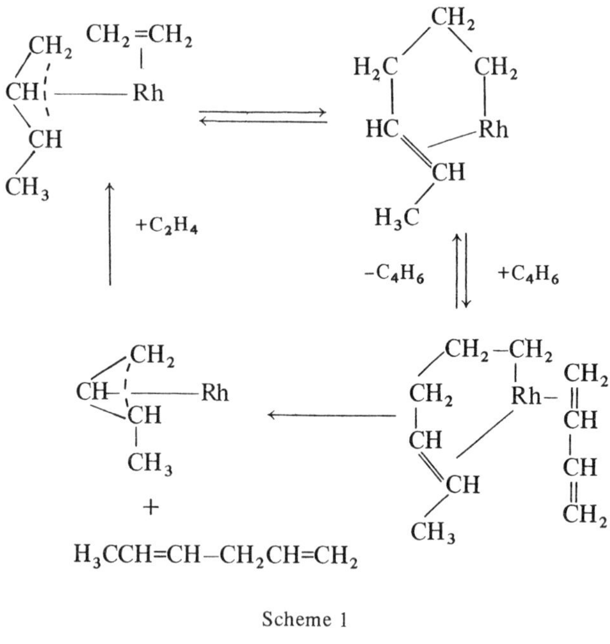

Scheme 1

In the reductive elimination reactions, the metal assumes a lower valence state. When a ligand which stabilizes the lower valence state is present, then the metal is not deactivated as it is by precipitation, and undergoes oxidative addition again. In order to stabilize the low valence state, it is necessary to remove the accumulated charge on the metal. For this purpose ligands which accept charges by back donation, such as triphenylphosphine, are useful. A complete theory of the role of the ligands has not yet been formulated. An approach to explain the role of the ligand in the coupling reaction of π-allyl complexes has been proposed by Traunmüller, Polansky, Heimbach, and Wilke and is based on molecular orbital calculations, taking the ionization potentials of the ligands into consideration [136].

In the foregoing, the formation of organic molecules on transition metal complexes is explained by stepwise processes of oxidative addition, insertion, and reductive elimination. One typical example, which can be clearly explained in this way, are the carbonylation and decarbonylation reactions catalyzed by rhodium complexes [10,137]. Tsuji and Ohno found that $RhCl(PPh_3)_3$ decarbonylates aldehydes and acyl halides under mild conditions stoichiometrically. Also this complex and $RhCl(CO)(PPh_3)_2$ are active for the catalytic decarbonylation at high temperature.

$$\mathrm{RCHO} + \mathrm{RhCl(PPh_3)_3} \longrightarrow \mathrm{RH} + \mathrm{RhCl(CO)(PPh_3)_2} + \mathrm{PPh_3}$$

$$\mathrm{RCOCl} + \mathrm{RhCl(PPh_3)_3} \longrightarrow \mathrm{RCORhCl_2(PPh_3)_2} \longrightarrow \mathrm{RCl} + \mathrm{RhCl(CO)(PPh_3)_2}$$

The reverse of the decarbonylation reaction, namely carbonylation can be catalyzed by the same complex.

$$\mathrm{CH_2{=}CH{-}CH_2Cl} + \mathrm{CO} \xrightarrow{\mathrm{RhCl(CO)(PPh_3)_2}} \mathrm{CH_2{=}CHCH_2COCl}$$

These stoichiometric and catalytic decarbonylation and carbonylation reactions can be explained by the following mechanism:

$$\mathrm{RhCl(CO)(PPh_3)_2} + \mathrm{RCO{-}X} \rightleftarrows \mathrm{RCORh(CO)Cl(PPh_3)_2X}$$

$$+\mathrm{CO} \uparrow\downarrow -\mathrm{CO}$$

$$\mathrm{RhCl(PPh_3)_3} + \mathrm{RCO{-}X} \longrightarrow \mathrm{RCORhClX(PPh_3)_2}$$

$$\uparrow\downarrow$$

$$\mathrm{RhCl(CO)(PPh_3)_2} + \mathrm{R{-}X} \rightleftarrows \mathrm{RRhClX(CO)(PPh_3)_2}$$

(or olefin)

The first step of the stoichiometric decarbonylation is the oxidative addition of acyl halide to $RhCl(PPh_3)_3$, which is d^8 complex, to form a five-coordinate d^6 acyl complex, which is still unsaturated. Then follows the acyl-alkyl rearrangement which gives a six-coordinated d^6 alkyl complex. The final step is a reductive elimination by the coupling of coordinated ligands or β-elimination to form $RhCl(CO)(PPh_3)_2$ and olefin or alkyl halide. The catalytic decarbonylation is initiated by the oxidative addition of an acyl halide to $RhCl(CO)(PPh_3)_2$ to form a six-coordinated acyl complex. Since the reaction temperature is high, elimination of the coordinated carbon monoxide takes place, which makes the acyl-alkyl rearrangement possible as a next step. The final step is a similar reductive elimination to regenerate $RhCl(CO)(PPh_3)_2$, thus completing the catalytic cycle. The carbonylation reaction can be explained as the exact reverse of the process of decarbonylation, and starts from the oxidative addition of alkyl halide to the rhodium complex. The decarbonylation reaction is a useful one, and several applications in organic synthesis have been reported. For example, deuterated alkane formation by the decarbonylation of the deuterated aldehyde was reported by Walborsky and Allen [138].

$$R-\underset{}{\overset{\displaystyle D}{\overset{|}{C}}}{=}O + RhCl(PPh_3)_3 \longrightarrow R-D + RhCl(CO)(PPh_3)_2 + PPh_3$$

It is generally believed that the cobalt-carbon σ-bond is rather unstable and reactive. However, *cobaloximes,* model complexes of vitamin B_{12}, form simple alkylcobaloximes by several ways, such as reductive alkylation, addition of olefins, and substitution reaction. The cobalt-carbon bonds thus formed are surprisingly stable, and undergo a variety of cleavage reactions, such as β-elimination, coupling, substitution with nucleophiles, and other reactions. The effect of this specific ligand is remarkable.

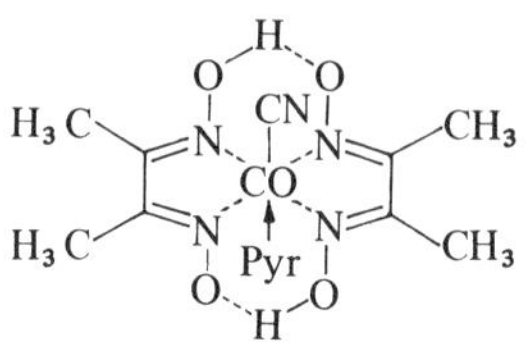

cyanopyridinocobaloxime

These reactions are fascinating, but difficult to survey briefly. A review was given by Schrauzer [139].

V. Cyclization Reaction and Related Reactions

One of the most characteristic reactions using transition metal complexes is the formation of various cyclic compounds from olefinic compounds. As described before, some cyclization reactions proceed via π-allyl and σ-bond formation and the coupling of the coordinated ligands. In the cycloaddition reactions, formation of a π-complex, rather than a σ-complex, sometimes plays an important role. In these reactions, the metal exhibits not only an electronic effect, but also a template effect, and the reactions are called *template ligand reactions.* In other words, the metals facilitate the mutual approach of the reactants and organize the reacting molecules in a suitable orientation. When two ligands are coordinated in adjacent positions on a suitable coordination sphere, the possibility of their reaction is greatly enhanced.

The best known example is the *cyclization* of butadiene and acetylene [127, 140]. Butadiene forms cyclooctadiene and cyclododecatriene by the catalytic action of nickel, iron, and other metal complexes. By an experiment using an iron complex with deuterated butadiene, it was proved that no hydrogen shift takes place in the cyclization reaction [70].

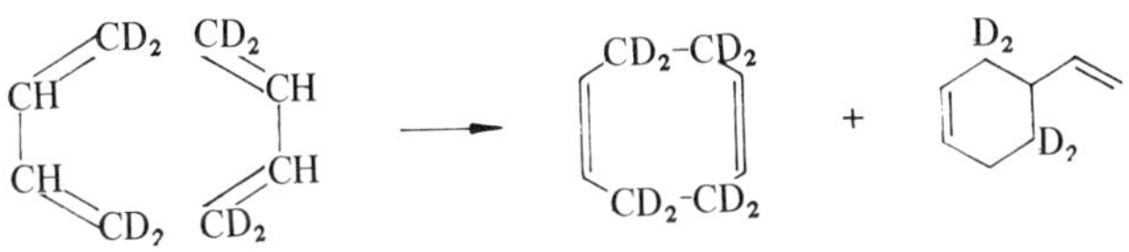

A recently reported cyclodimerization reaction of butadiene is the formation of 2-methylenevinylcyclopentane [170,171]. The reaction is catalyzed most effectively by *trans*-bis (triethylphosphine)chloro-(o-tolyl)nickel in the presence of alcohol. It is interesting that negligible amounts of cyclooctadiene and cyclododecatriene are formed with this catalyst.

Several mechanisms have been proposed for the cyclization of acetylene to *benzene derivatives.* Whitesides and Ehrmann proved the concerted cyclotrimerization mechanism by carrying out the cyclization reaction with $CD_3C{=}CCH_3$ catalyzed by Cr, Co, Ni, and Ti complexes and analyzing the distribution of D in the reaction products [141].

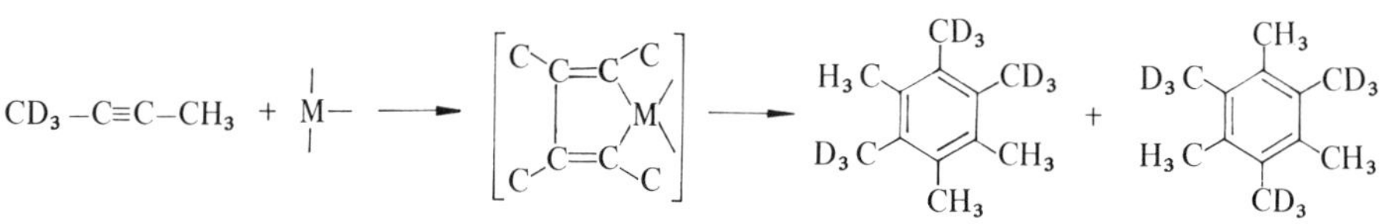

Further work on the conversion of triphenylchromium to tetramethylnaphthalene on reaction with 2-butyne was carried out by Whitesides and Ehrmann, and the reaction was explained by insertion of 2-butyne in a phenylchromium σ-

bond to form a styrylchromium reagent, cyclization of the latter to a benzchromole, and reaction of the benzchromole with a second molecule of 2-butyne to form tetramethylnaphthalene [142].

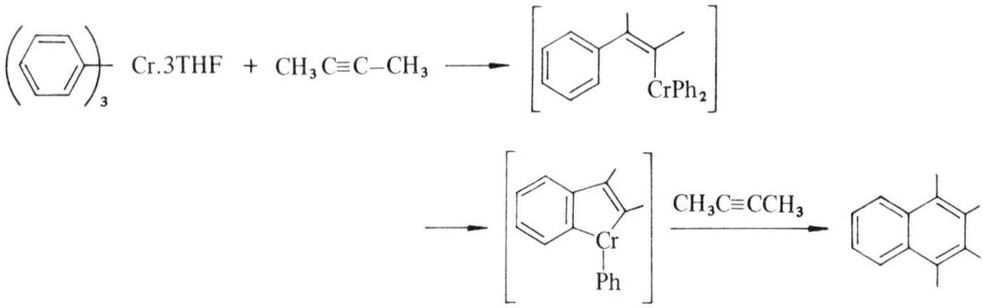

It is well known that palladium chloride is an active catalyst for the cyclization of acetylene to form cyclobutadiene as well as benzene derivatives. In this reaction an intermediate complex was isolated which has a palladium carbon σ-bond, the formation of which was explained by an insertion mechanism, not by concerted cyclotrimerization. When this complex obtained from butyne and palladium chloride was decomposed by various means, 5-vinyl-1,2,3,4,5-pentamethylcyclopentadiene and 5-(1-chlorovinyl)-1,2,3,4,5-pentamethylcyclopentadiene were obtained in addition to hexamethylbenzene [143,144].

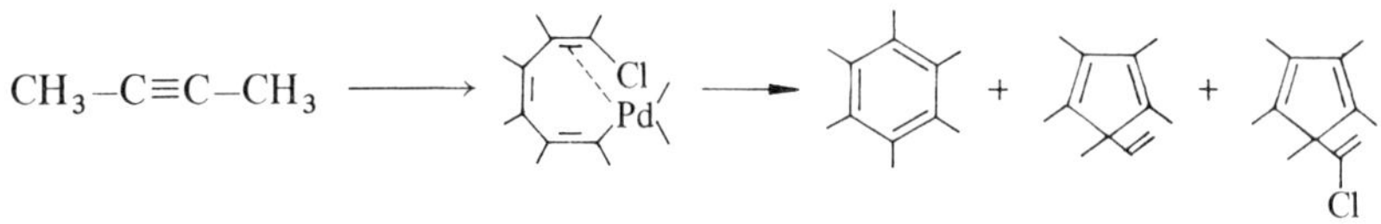

A Diels-Alder type reaction can be catalyzed with metal complexes. For example, $Fe(COT)_2$ can be used as the catalyst for the addition reaction of butadiene and acetylene [145]. In the absence of the catalyst, this reaction is not possible.

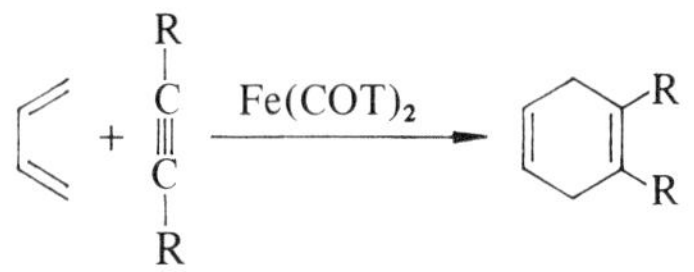

COT: cyclooctatetraene

With nickel catalyst, butadiene and acetylene react to form 1,4,7-cyclodecatriene [146,147].

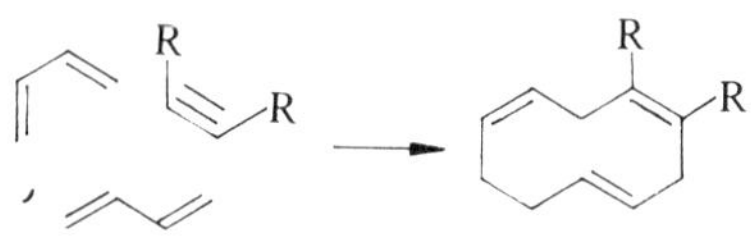

The $Fe(COT)_2$ complex is also an active catalyst for the reaction of acetylene and norbornadiene to form benzene derivatives and cyclopentadiene [148,149]. In this reaction, too, norbornadiene and two molecules of acetylene coordinate to the catalyst and cyclization takes place to liberate cyclopentadiene.

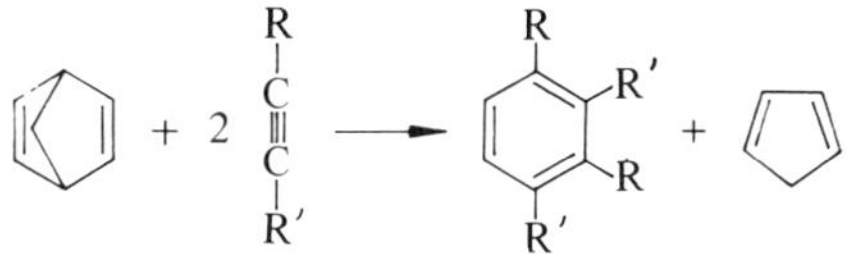

The stereospecific dimerization of norbornadiene catalyzed by a dimeric cobalt complex, is called a *π-complex multicenter reaction* [150–153].

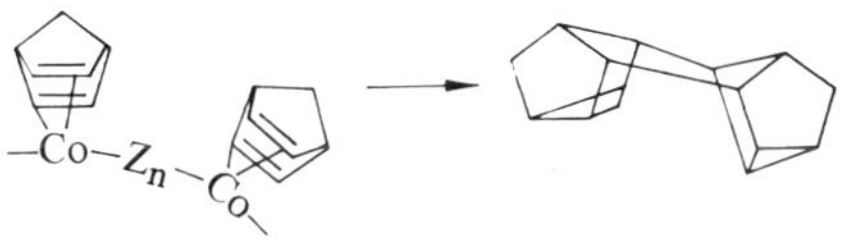

Addition of cyclopropane to activated olefins to form cyclopentanes is catalyzed by a zero-valent nickel complex [154]. Thus methylenecyclopropane and methyl acrylate react in the presence of a catalytic amount of bis(acrylonitrile nickel) to give methyl 3-methylenecyclopentanecarboxylate in 82% yield.

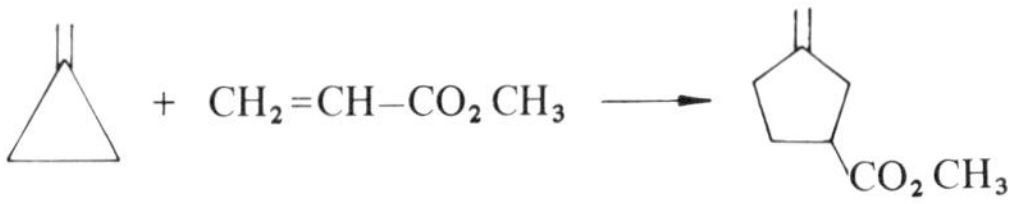

A case in which π-coordination of olefins to metallic complexes plays a decisive part is the remarkable reaction called *olefin metathesis*. Calderon, Chen, and Scott reported the reaction in which two olefinic bonds are broken and two new olefinic bonds are formed via a four-membered ring [155,156]. For example, 2-pentene is converted into 3-hexene and 2-butene.

$$C{-}C{=}C{-}C{-}C \rightleftarrows C{-}C{=}C{-}C + C{-}C{-}C{=}C{-}C{-}C$$

This reaction can be expressed as follows:

$$\underset{50\%}{2R{-}CH{=}CH{-}R'} \rightleftarrows \underset{25\%}{R{-}CH{=}CH{-}R} + \underset{25\%}{R'CH{=}CH{-}R'}$$

The catalysts used in this reaction are $WCl_6/AlEtCl_2/EtOH$, $WCl_6/BuLi$. Some Mo complexes can be used similarly [158,159]. The mechanism of the reaction, namely the formation of a four-membered ring, was proved by experiments using deuterium. Thus, in the metathesis reaction of 2-butene and 2-butene-d_8, the newly formed 2-butene has mass number 60, and no butene of mass 59 or 61 was formed [156].

Some examples of the usefulness of this metathesis reaction from the standpoint of organic synthesis are cited in the following. Conversion of inner olefins into terminal olefins would be very useful. However, it is thermodynamically "uphill" and can only occur if a second reagent is consumed concurrently, for example, by hydroboration.

The conversion can be done by metathesis in the form of the *"ethenolysis" reaction.* In this, inner olefins are treated with ethylene to form two moles of terminal olefins. For example, by the reaction of

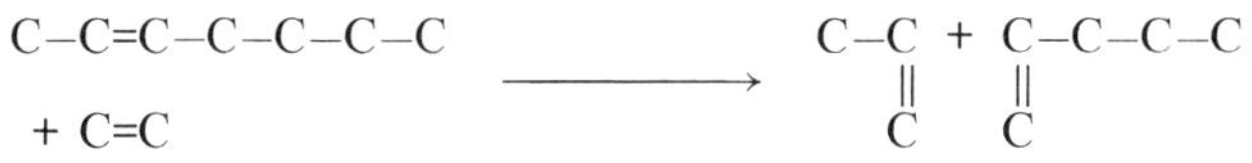

ethylene and 2-heptene, a mixture of olefins consisting of propylene (6%), 2-butene (3%), 1-hexene (19%), 2-heptene (66%), and 5-decene (7%) was obtained. The reaction is not selective and the following reaction proceeds at the same time to give 2-butene and 5-decene.

The application of the metathesis reaction to cyclic olefins affords large-membered cyclic olefins [160].

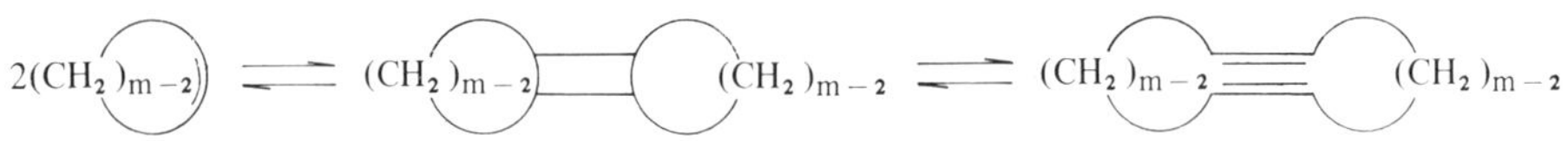

Of course, in this reaction of cyclic olefins, the reaction does not necessarily stop at the first step, but proceeds further to form trimer, tetramer, and polymers by sequential metathesis reactions. For example, Wideman carried out the metathesis of cyclooctene and obtained 1,9-cyclohexadecadiene in 20% yield, from which 16-membered monoketone and diketone were synthesized [161].

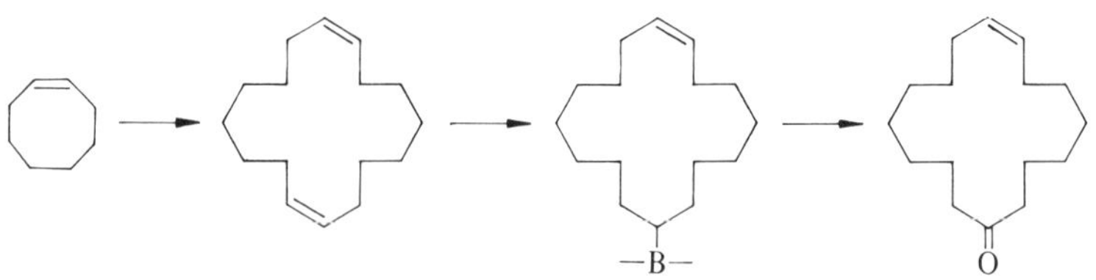

The mechanism of the metathesis reaction is explained by the following three steps, as proposed by Calderon and others.

1. Bisolefin-metal complex formation

2. Transalkylidenation

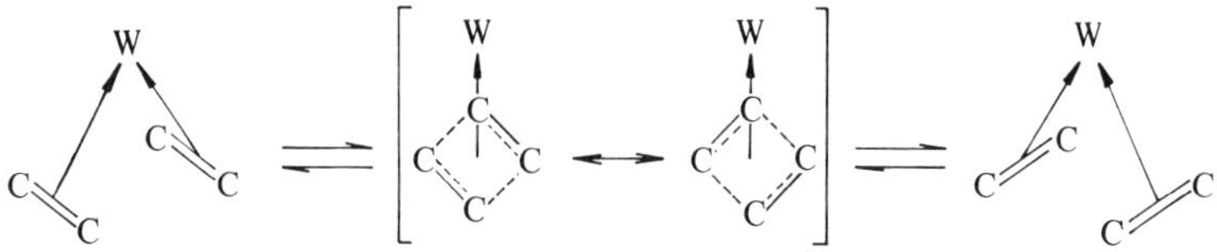

3. Olefin exchange

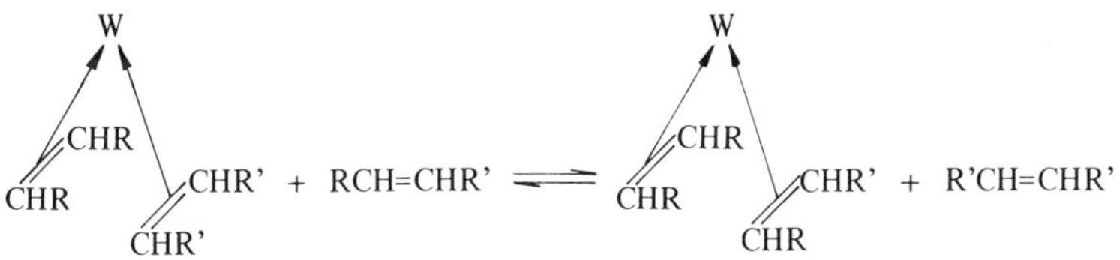

Tungsten requires two olefinic ligands in a *cis*-configuration, and then an electronically excited four membered-ring is formed.

On the basis of the concept of *molecular orbital symmetry conservation* proposed by Woodward and Hoffmann [173)], Mango and Schachtschneider demonstrated that a reaction pathway does exist for orbital symmetry conservation if the two olefinic ligands are coordinated to the transition metal in a four-membered ring [162-164)]. The formation of cyclobutane by the cycloaddition reaction of two olefinic bonds is forbidden in the absence of the transition metal catalyst. The role of the catalyst is the *switching of symmetry-forbidden to symmetry-allowed* through an appropriate manipulation of bonding and non-bonding electrons. The catalyst behaves as an electron relay switch and need not undergo significant charge generation or electron excitation. This consideration gives some insight into the mysterious role of transition metal catalysts. Vand der Lugt has proposed a somewhat different explanation of the reaction [165)].

An interesting *cocyclization reaction of butadiene with aldehydes catalyzed by palladium* was reported independently by three groups [47,48,166)]. In the presence of a palladium triphenylphosphine complex as catalyst, butadiene and aldehyde give 2-substituted 3,6-divinyltetrahydropyran and 1-substituted 2-vinyl-4,6-heptadien-1-ol. Interestingly, the latter formed when the ratio of PPh_3/Pd in the catalyst was about one, while the former was the main product when the ratio was larger than two. The mechanism proposed by Ohno and Tsuji for this new cyclization reaction is the following [48].

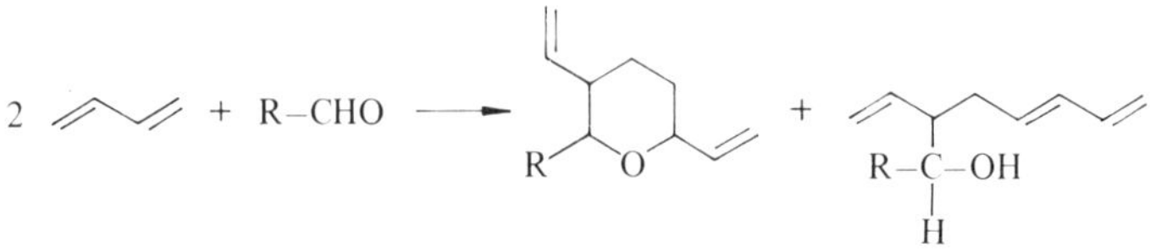

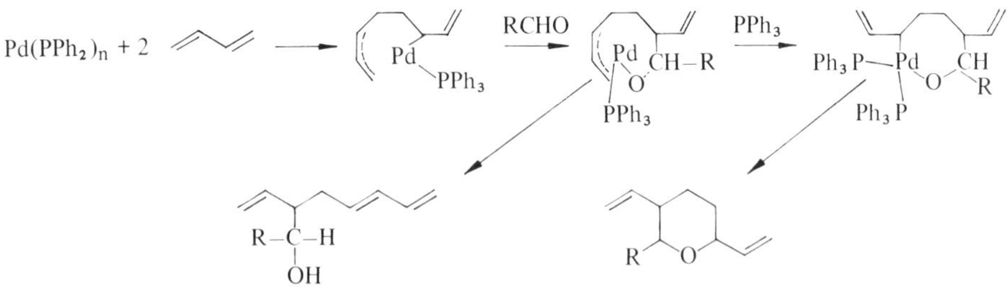

This type of cyclization of butadiene with heteropolar double bonds seems to be a general reaction. Tsuji and Ohno reported that isocyanates react with conjugate diene to form 1-substituted 3,6-divinylpiperidone. For example, isoprene and phenyl isocyanate gave *trans* and *cis*-3,6-diisopropenyl-1-phenyl-2-piperidone in the presence of palladium phosphine complexes [167].

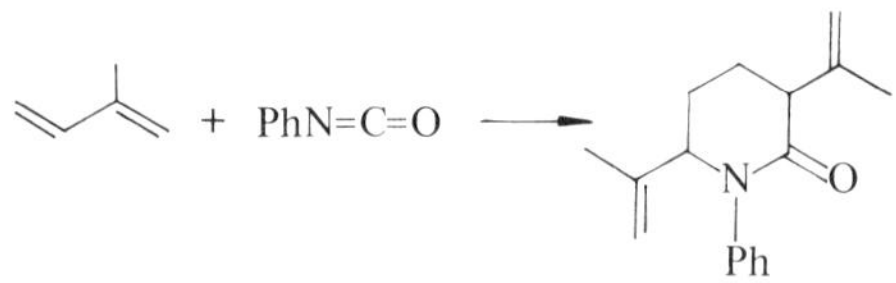

VI. Concluding Remarks

The general patterns of organic syntheses involving transition metal complexes have been discussed with some typical examples. It should be said that studies of transition metal catalysis are still in a relatively early stage. Much has to be done before we can understand what is really happening during the formation of organic compounds in the coordination spheres of transition metals. In this sense, we may expect that unexpected and exciting new reactions will be found in the course of further careful studies of transition metal complexes.

Acknowledgments. The author expresses his sincere appreciation to Dr. H. G. Tennent (Hercules Incorp.) for his careful reading of the manuscript and for suggestions. Useful suggestions by Dr. M. Ryang (Osaka Univ.) and Dr. A. Yamamoto (Tokyo Inst. Tech.) are gratefully acknowledged.

VII. References

1) Parshall, G. W., Mrowca, J. J.: Advan. Organometal. Chem. *7,* 157 (1968).
2) Collman, J. P.: Accounts Chem. Res. *1,* 136 (1968).
3) – Roper, W. R.: Advan. Organometal. Chem. *7,* 54 (1968).
4) Osborn, J. A., Jardine, F. H., Young, J. F., Wilkinson, G.: J. Chem. Soc. *A,* 1711 (1966).
5) Gustorf, E. K., Grevels, F. W., Hogan, J. C.: Angew. Chem. *81,* 918 (1969).
6) Malatesta, L., Cariello, C.: J. Chem. Soc. *1958,* 2323.
7) Ugo, R., Cariati, F., LaMonica, G.: Chem. Commun. *1966,* 868.
8) Vaska, L.: Account Chem. Res. *1,* 335 (1968).
9) Tsuji, J., Ohno, K.: J. Am. Chem. Soc. *90,* 94 (1968).
10) Ohno, K., Tsuji, J.: J. Am. Chem. Soc. *90,* 99 (1968).
11) Harvie, I., Kemmitt, R. D. W.: Chem. Commun. *1970,* 198.
12) Tripathy, P. B., Roundhill, D. M.: J. Am. Chem. Soc. *92,* 3825 (1970).
13) Collman, J. P., Kang, J. W.: J. Am. Chem. Soc. *89,* 844 (1967).
13a) Roundhill, D. M., Jonassen, H. B.: Chem. Commun. *1968,* 1233.
14) Bennett, M. A., Milner, D. L.: Chem. Commun. *1967,* 581 and J. Am. Chem. Soc. *91,* 6983 (1969).
15) Levison, J. J., Robinson, S. D.: J. Chem. Soc. *A,* 639 (1970).
16) Parshall, G. W., Knoth, W. H., Schunn, R. A.: J. Am. Chem. Soc. *91,* 4990 (1969).
17) Fleischer, E. B., Lavallee, D.: J. Am. Chem. Soc. *89,* 7132 (1967).
18) Chatt, J., Davidson, J. M.: J. Chem. Soc. *1965,* 843.
19) Saito, T., Uchida, Y., Misono, A., Yamamoto, A., Morifuji, K., Ikeda, S.: J. Organometal. Chem. *6,* 572 (1966).
20) Roundhill, R. D.: Inorg. Chem. *9,* 255 (1970) and Chem. Commun. *1969,* 567.
21) Chatt, J., Eaborn, C., Kapoor, P. N.: J. Chem. Soc. *A,* 881 (1970).
22) Fitton, P., McKeon, J. E.: Chem. Commun. *1968,* 4.
23) Bland, W. J., Kemmitt, R. D.: J. Chem. Soc. *A,* 1278 (1968).
24) Fahey, D. R.: J. Am. Chem. Soc. *92,* 402 (1970).
25) Jones, K., Wilke, G.: Angew. Chem. *81,* 534 (1969).
26) Schott, H., Wilke, G.: Angew. Chem. *81,* 896 (1969).
27) Dawans, F., Marechal, J. C., Teyssis, PH.: J. Organometal. Chem. *21,* 259 (1970).
28) Chiusoli, G. P.: Angew. Chem. (Intern. Ed.) *6,* 124 (1967).
29) Bauld, N. C.: Tetrahedron Letters *1963,* 1841.
30) Rhee, I., Mizuta, N., Ryang, M., Tsutsumi, S.: Bull. Chem. Soc. Japan *41,* 1417 (1968).
31) Tsutsumi, S., Yoshisato, E.: J. Am. Chem. Soc. *90,* 4488 (1968) and Chem. Commun. *1968,* 33.
32) Corey, E. J., Hegedus, L. S.: J. Am. Chem. Soc. *91,* 1233 (1969).
33) Coffey, C. E.: J. Am. Chem. Soc. *83,* 1623 (1961).
34) Emerson, G. F., Watts, L., Pettit, R.: J. Am. Chem. Soc. *87,* 131 (1965).
35) Gowling, E. W., Kettle, S. F. A., Scharples, G. M.: Chem. Commun. *1968,* 21.
36) Dubini, M., Montino, F., Chiusoli, G. P.: Chim. Ind. (Milan) *49,* 839 (1965).
37) Corey, E. J., Semmelhack, M. F.: J. Am. Chem. Soc. *89,* 2755 (1967).
38) Heimbach, P., Jolly, P. W., Wilke, G.: Advan. Organometal. Chem. *8,* 29 (1970).
39) Tsuji, J., Morikawa, M., Takahashi, H.: Tetrahedron Letters *1965,* 4387.
40) – Takhashi, H., Morikawa, M.: Kogyo Kagaku Zasshi *69,* 920 (1966).
41) Takahashi, S., Hagihara, N.: Tetrahedron Letters *1967,* 2451.
42) – Shibano, T., Hagihara, N.: Bull. Chem. Soc. Japan *41,* 254 (1968).
43) – – – Bull. Chem. Soc. Japan *41,* 454 (1968).

44) – – – Chem. Commun. *1969,* 161.
45) Smutny, E. J.: J. Am. Chem. Soc. *89,* 6793 (1967).
46) Hata, G., Takahashi, K., Miyake, A.: Chem. Ind. 1836 (1969).
47) Manyik, R. M., Walker, W. E., Atkins, K. E., Hammack, E. S.: Tetrahedron Letters *1970,* 3813.
48) Ohno, K., Mitsuyasu, T., Tsuji, J.: Tetrahedron Letters *1971.* 67.
49) Heimbach, P.: Angew. Chem. *80,* 967 (1968).
50) Hieber, W., Lindner, E.: Chem. Ber. *94,* 1417 (1961).
51) – Duchatsh, H.: Chem. Ber. *98,* 2933 (1965).
52) Otsuka, S., Rossi, M.: J. Chem. Soc. *A,* 497 (1969).
53) Slaugh, L. H., Mullineaux, R. D.: J. Organometal. Chem. *13,* 469 (1968).
54) Kniese, W., Nienberg, H. J., Fischer, R.: J. Organometal. Chem. *17,* 133 (1969).
55) Tucci, E. R.: Ind. Eng. Chem. *7,* 32 (1968) and *8,* 286 (1969).
56) Imjanitov, N. S., von, Rudkovski, D. M.: J. Prakt. Chem. *311,* 712 (1969).
57) Chalk, A. J., Harrod, J. F.: J. Am. Chem. Soc. *89,* 1640 (1967).
58) Susuki, T., Tsuji, J.: Tetrahedron Letters *1968,* 913.
59) – – J. Org. Chem. *35,* 2982 (1970).
60) Bamford, C. H., Eastmond, G. C., Writtle, D.: J. Organometal. Chem. *17,* 33 (1969). and the references cited therein.
61) Mori, Y., Tsuji, J.: Tetrahedron *27,* 3811 (1971).
62) – – Tetrahedron *27,* 4039 (1971).
63) Abel, E. W., Keppie, S. A., Lappert, M. F., Moorhouse, S.: J. Organometal. Chem. *22,* C31 (1970).
64) Jarchow, O., Schultz, H., Nast, R.: Angew. Chem. *82,* 43 (1970).
65) Hashimoto, I., Ryang, M., Tsutsumi, S.: Tetrahedron Letters *1969,* 3291.
65a) – Tsuruta, N., Ryang, M., Tsutsumi, S.: J. Org. Chem. *35,* 3748 (1970).
66) – Ryang, M., Tsutsumi, S.: Tetrahedron Letters *1970,* 4567.
67) Cross, R.: J. Organometal. Chem. Rev. *2,* 97 (1967).
68) Yamamoto, A., Morifuji, K., Ikeda, S., Saito, T., Uchida, Y., Misono, A.: J. Am. Chem. Soc. *87,* 4652 (1965).
69) Saito, T., Uchida, Y., Misono, A., Yamamoto, A., Morifuji, K., Ikeda, S.: J. Am. Chem. Soc. *88,* 5198 (1966).
70) – – – – – – J. Organometal. Chem. *6,* 572 (1966).
71) Yamamoto, A., Morifuji, K., Ikeda, S., Saito, T., Uchida, Y., Misono, A.: J. Am. Chem. Soc. *90,* 1878 (1968).
72) Yagupsky, G., Mowat, W., Shortland, A., Wilkinson, G.: Chem. Commun. *1970,* 1369.
73) Cross, R. J., Wardle, R.: J. Chem. Soc. *A,* 841 (1970).
74) Heck, R. F.: J. Am. Chem. Soc. *90,* 5518 (1968).
75) – J. Am. Chem. Soc. *90,* 5526 (1968).
76) – J. Am. Chem. Soc. *90,* 5531 (1968).
77) – J. Am. Chem. Soc. *90,* 5535 (1968).
78) – J. Am. Chem. Soc. *90,* 5538 (1968).
79) – J. Am. Chem. Soc. *90,* 5542 (1968).
80) – J. Am. Chem. Soc. *90,* 5546 (1968).
81) – J. Am. Chem. Soc. *91,* 6707 (1969).
82) Henry, P. M.: Tetrahedron Letters *1968,* 2285.
83) Seyferth, D., Spohn, R. J.: J. Am. Chem. Soc. *90,* 540 (1968).
84) – – J. Am. Chem. Soc. *91,* 3037 (1969).
85) Ryang, M., Song, K. M., Sawa, Y., Tsutsumi, S.: J. Organometal. Chem. *5,* 305 (1966).
86) Fischer, E. O., Maasböl, A.: Chem. Ber. *100,* 2445 (1967).
87) Ryang, N.: Organometal. Chem. Revs. *A,* 5, 67 (1970).

88) Seebach, D.: Angew. Chem. (Intern. Ed.) *8,* 639 (1969).
89) Ryang, M., Rhee, E., Tsutsumi, S.: Bull. Chem. Soc. Japan *37,* 341 (1964) and *38,* 330 (1965).
90) Fischer, E. O., Maasbol, A.: Ger. Pat. 1214233 (1966).
91) Sawa, Y., Ryang, M., Tsutsumi, S.: Tetrahedron Letters *1969,* 5189.
92) — — — Abstracts, Symposium on Organometal. Chem. Japan *1970,* 138.
93) Corey, E. J., Hegedus, L. S.: J. Am. Chem. Soc. *91,* 4926 (1969).
94) Sawa, Y., Hashimoto, I., Ryang, M., Tsutsumi, S.: J. Org. Chem. *33,* 2159 (1968).
95) Fukuoka, S., Ryang, M., Tsutsumi, S.: J. Org. Chem. *33,* 2973 (1968).
96) — — — Unpublished results. Private Commun.
97) Cooke, M. P.: J. Am. Chem. Soc. *92,* 6080 (1970).
98) Fischer, E. O., Maasböl, A.: Angew. Chem. *76,* 645 (1964).
99) — Knauss, L.: Chem. Ber. *103,* 1262 (1970), and many other papers cited therein.
100) — Maasböl, A.: J. Organometal. Chem. *12,* P15 (1968).
101) — Dotz, K. H.: Chem. Ber. *103,* 1273 (1970).
102) Heck, R. F.: Accounts Chem. Res. *2,* 10 (1969).
103) Tsuji, J.: Accounts Chem. Res. *2,* 144 (1969).
104) — Advan. Org. Chem. *6,* 109 (1969).
105) Hüttel, R.: Synthesis *1970,* 225.
106) Aguilo, A.: Advan. Organometal. Chem. *5,* 321 (1967).
107) Stern, E. W.: Catalysis Revs. *1,* 73 (1967).
108) Takahashi, H., Tsuji, J.: J. Am. Chem. Soc. *90,* 2387 (1968).
109) Parshall, G. W.: Accounts Chem. Res. *3,* 139 (1970).
110) Moritani, I., Fujiwara, Y.: Tetrahedron Letters *1967,* 1119. – Fujiara, Y., Martani, I., Ikegami, K., Tanaka, R., Teranishi, S.: Bull. Chem. Soc. Japan *43,* 863 (1970) and references cited therein.
111) Takahashi, H., Tsuji, J.: J. Organometal. Chem. *10,* 511 (1967).
112) Fahey, D. R.: Chem. Commun. *1970,* 417.
113) Yamamoto, Y., Yamazaki, H., Hagihara, N.: J. Organometal. Chem. *18,* 189 (1969).
114) Otsuka, S., Nakamura, A., Yoshida, T.: J. Am. Chem. Soc. *91,* 7196 (1969).
115) Kajimoto, T., Tsuji, J.: J. Organometal. Chem. *23,* 275 (1970).
116) Heck, R. F.: J. Am. Chem. Soc. *85,* 3383 (1963).
117) Klein, H. S.: Chem. Commun. *1968,* 377.
118) O'Brien, S.: J. Chem. Soc. *A,* 9 (1970).
119) Pu, L. S., Yamamoto, A., Ikeda, S.: J. Am. Chem. Soc. *90,* 3896 (1968).
120) Misono, A., Uchido, Y., Hidai, M., Kuse, T.: Chem. Commun. *1968,* 981.
121) Haynes, P., Slaugh, L. H., Kohnle, J. F.: Tetrahedron Letters *1970,* 365.
122) Iwashita, Y., Hayata, A.: J. Am. Chem. Soc. *91,* 2525 (1969).
123) Halpern, J.: Ann. Rev. Phys. Chem. *16,* 103 (1965).
124) Moritani, I., Yamamoto, Y., Konishi, H.: Chem. Commun. *1969,* 1457.
125) Chiusoli, G. P., Bottaccio, G.: Chim. Ind. (Milan) *47,* 165 (1965).
126) Cassar, L., Chiusoli, G. P.: Chim. Ind. (Milan) *48,* 323 (1966).
127) Wilke, G., Kröner, M., Bogdanovic, B.: Angew. Chem. *73,* 755 (1961).
128) Breil, H., Wilke, G.: Angew. Chem. *82,* 355 (1970).
129) Corey, E. J., Wat, E. K. W.: J. Am. Chem. Soc. *89,* 2755 (1967).
130) — Hamanaka, E.: J. Am. Chem. Soc. *89,* 2758 (1967).
131) Takahashi, Y., Sakai, S., Ishii, Y.: Chem. Commun. *1967,* 1092.
132) Yamamoto, T., Yamamoto, A., Ikeda, S.: J. Am. Chem. Soc. *93,* 3350 (1971).
133) — Ikeda, S.: J. Am. Chem. Soc. *89,* 5989 (1967).
134) Tsutsui, M., Hancoch, M., Ariyoshi, J., Levy, M. N.: Angew. Chem. (Intern. Ed.) *8,* 410 (1969).

135) Cramer, R.: Accounts Chem. Res. *1,* 186 (1968).
136) Traunmüller, R., Polansky, O. E., Heimbach, P., Wilke, G.: Chem. Phys. Letters *3,* 300 (1969).
137) Tsuji, J., Ohno, K.: Synthesis *1969,* 157.
138) Walborsky, H. M., Allen, L. B.: Tetrahedron Letters *1970,* 823.
139) Schrauzer, G. N.: Accounts Chem. Res. *1,* 97 (1968).
140) Reppe, W., Kutepow, N. V., Magin, A.: Angew. Chem. *81,* 717 (1969).
141) Whitesides, G. M., Ehrmann, W. J.: J. Am. Chem. Soc. *91,* 3800 (1969).
142) – – J. Am. Chem. Soc. *92,* 5625 (1970).
143) Dietl, H., Reinheimer, H., Moffat, J., Maitlis, P. M.: J. Am. Chem. Soc. *90,* 2276 (1970).
144) Reinheimer, H., Moffat, J., Maitlis, P. M.: J. Am. Chem. Soc. *90,* 2285 (1970).
145) Carbonaro, A., Greco, A., Dall'Asta, G.: J. Org. Chem. *33,* 3948 (1968).
146) Brenner, W., Heimbach, P., Wilke, G.: Angew. Chem. *78,* 983 (1966).
147) – – – Ann. Chem. *727,* 194 (1969).
148) Carbbonaro, A., Greco, A., Dall'Asta, G.: Tetrahedron Letters *1968,* 5129.
149) – – – J. Organometal. Chem. *20,* 177 (1969).
150) Schrauzer, G. N., Bastian, B. N., Fosselius, G. A.: J. Am. Chem. Soc. *88,* 4980 (1966).
151) – Advan. Catalysis *18,* 373 (1968).
152) – Ho, R. K. Y., Schiesinger, G.: Tetrahedron Letters *1970,* 543.
153) Boer, F. P., Tsai, J. H., Flynn, J. J.: J. Am. Chem. Soc. *92,* 6093 (1970).
154) Noyori, R., Odagi, T., Takaya, H.: J. Am. Chem. Soc. *92,* 5781 (1970).
155) Calderon, N., Chen, H. Y., Scott, K. W.: Tetrahedron Letters *1967,* 3327.
156) – Ofstead, E. A., Ward, J. P., Judy, W. A., Scott, K. W.: J. Am. Chem. Soc. *90,* 4133 (1968).
157) Wang, J., Menapace, H. R.: J. Org. Chem. *33,* 3794 (1968).
158) Zuech, E. A.: Chem. Commun. *1968,* 1182.
159) – Hughes, W. B., Kubicek, D. H., Kittleman, E. T.: J. Am. Chem. Soc. *92,* 528 (1970).
160) Wasserman, W., BenFraim, D. A., Wolovsky, R.: J. Am. Chem. Soc. *90,* 3286 (1968).
161) Wideman, L. G.: J. Org. Chem. *33,* 4541 (1968).
162) Mango, F. D., Schachtschneider, J. H.: J. Am. Chem. Soc. *89,* 2484 (1967); – *91,* 1030 (1969).
163) – Tetrahedron Letters *1969,* 4813.
164) – Advan. Catalysis *20,* 291 (1969).
165) Vand der Lugt, Th. A. M.: Tetrahedron Letters *1970,* 2281.
166) Haynes, P.: Tetrahedron Letters *1970,* 3687.
167) Ohno, K., Tsuji, J.: Chem. Commun. *1971,* 247.
168) Sneedon, R. P. A., Zeiss, H. H.: J. Organometal. Chem. *22,* 713 (1970).
169) Light, J. R. C., Zeiss, H. H.: J. Organometal. Chem. *21,* 517 (1970).
170) Kiji, J., Masui, K., Furukawa, J.: Tetrahedron Letters *1970,* 2561.
171) – – – Chem. Commun. *1970,* 1300.
172) Murahashi, S., Horiie, S.: J. Am. Chem. Soc. *78,* 4816 (1956).
173) Hoffmann, R., Woodward, R. B.: Accounts Chem. Res. *1,* 17 (1968).

Received January 4, 1971

The Stability and Structures of Olefin and Acetylene Complexes of Transition Metals

Dr. Leslie D. Pettit
Department of Inorganic and Structural Chemistry, The University, Leeds, U.K.

Dr. David S. Barnes
Department of Chemistry, The University, Keele, Staffs., U.K.

Contents

A. Introduction

Olefin and acetylene complexes have been known for many years, the first, Zeise's salt, $K[PtCl_3.C_2H_4]$, being isolated 140 years ago. These early complexes were predominantly of curiosity value only and only in the last 20 years have detailed studies been carried out on synthetic techniques, kinetics and mechanisms of reactions, stabilities and structures.

The impetus for research in this filed stems from the industrial importance of metal-olefin complexes as intermediates and catalysts in a wide range of reactions, especially in the petrochemical industry. Major uses include the Wacker Process (oxidation of ethylene to acetaldehyde in the presence of $PdCl_2$), the OXO process (hydroformalation of olefins), the specific hydrogenation of double bonds and the isomerisation of olefins (e.g. but-1-ene to but-2-ene in the presence of $[(C_2H_4)_2RhCl]_2$).

Studies on the preparation and reactions of olefin complexes have been the subject of several recent reviews. In particular olefin and acetylene complexes of platinum and palladium have been reviewed by Hartley [1], characterised olefin-metal complexes by Quinn and Tsai [2] and some acetylene-metal complexes by Greaves, Lock and Maitlis [3].

Olefins and acetylenes can, potentially, bond to metal as either *pi* or σ ligands. In the former case the metal atom is approximately equidistant from two carbon atoms linked by a multiple bond while in the latter case the metal atom is bonded to one particular carbon atom by a covalent σ-bond.

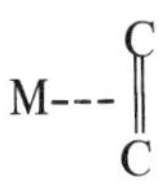

a *pi*-complex
or μ-complex

M—C=C

a σ-complex

In this review the word *pi* will be used for identifying the type of complex often termed a μ-complex, and the symbol π will be restricted to the π-bond having an angular momentum of 1 unit about the internuclear axis.

The *pi*-complex formed by an unsaturated ligand can involve bonding between the metal and two carbon atoms linked by a double bond or bonding to three or more carbon atoms which are part of an aromatic conjugated system. The former class will include Zeise's Salt (A. I) and complexes of cyclooctadiene (COD) in which the two olefin bonds are not conjugated (A. II), while the latter class will include *pi*-allyl and butadiene complexes and metal arenes (A. III–V). In this review the structures and stabilities of complexes of *pi*-bonded olefins and acetylene which are not part of a conjugated system will be considered, allylic complexes and complexes of more extensive conjugated systems being omitted.

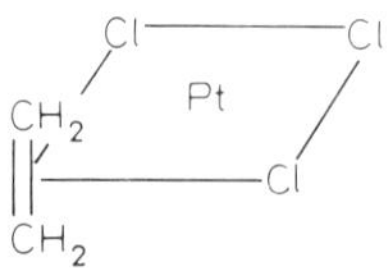

A.I

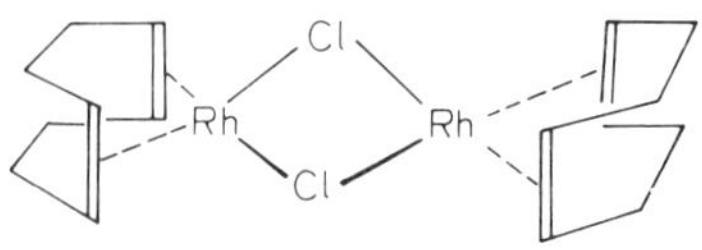

A.II

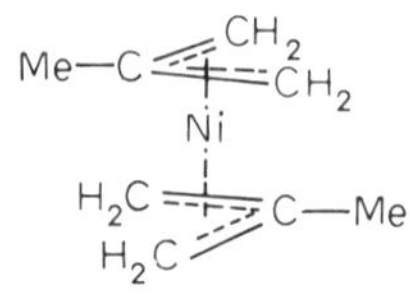

A.III

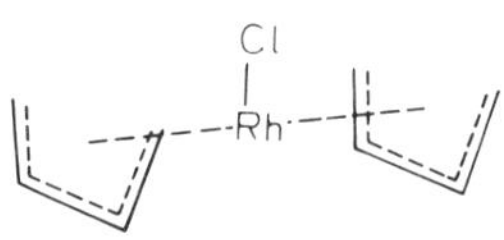

A.IV

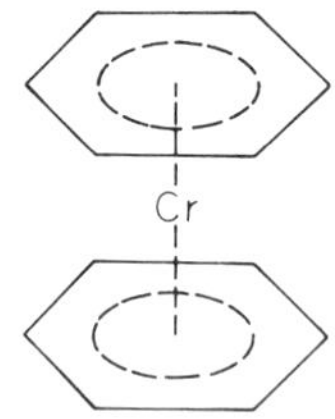

A.V

B. Types of Olefin and Acetylene Complexes

1. Olefin Complexes

i) Monodentate – Most frequently the olefinic bond can act as a monodentate ligand, as in Zeise's Salt (A. I) or COD complexes (A. II). The ethylene molecule is oriented approximately at right angles to the plane of the other co-ordinated ligands, taking up one co-ordination position. The resulting complex may contain more than one co-ordinated olefin (*e.g.* $[(C_2H_4)_2RhCl]_2$).

ii) Bidentate – In some cases the olefinic bond can act as a pseudo-bidentate ligand co-ordinating in the plane of the complex. This type of co-ordination appears to be stabilized by metals in low oxidation states and ligands containing strongly electron withdrawing substituents, *e.g.* $[Pt(P\phi_3)_2.C_2(CN)_4]$ [4)] shown in D. XII.

iii) *pi*-σ-Ligands – Suitable ligands can act as *pi* and σ-ligands simultaneously, *e.g.* allylchloroplatinum (II) is a tetramer with the structure shown in D. XVIII. This form of co-ordination is found more with platinum than palladium, which tends to form normal *pi*-allyl complexes.

iv) Allene complexes – Allenes ($R_2C{=}C{=}CR'_2$) act as monodentate ligands to a particular metal atom although the two olefin bonds can each co-ordinate separately to different metal atoms as in the dirhodium complex shown in D. XVI [6)]. Allenes can also form σ-bonded complexes.

2. Acetylene Complexes

i) Monodentate – Monodentate complexes are comparable to the monodentate olefin complexes, *e.g.* $[(Bu^t.C{\equiv}C.Bu^t)$ (*p*-toluidine)$PtCl_2]$ has the structure shown in D. XXII, the triple bond being at right angles to the plane of the complex [7].

ii) Bidentate – As with olefins, acetylenes appear to act as bidentate ligands in zerovalent metal complexes. From n.m.r. spectra complexes of formula $L_2M(RC{\equiv}CR')$ appear to be approximately planar in solution [3] and X-ray studies show the complex $[Pt(PPh_3)_2(PhC{\equiv}CPh)]$ to be almost planar (D. XXIII).

iii) Bridging – The acetylenic bond can bridge two metal atoms, each π-orbital apparently behaving as separate olefinic π-bonds at right angles. $[Co_2(CO)_6(PhC{\equiv}CPh)]$ has the structure shown in D. XXIV [8].

iv) *pi*-σ-Ligands – The complex $[Co_4(CO)_{10}(EtC{\equiv}CEt)]$ shown in D. XXV has two σ-bonds and a multicentre *pi*-bond between two cobalt atoms and the multiple bond [9].

C. Stability of Olefin and Acetylene Complexes

1. Introduction

Whilst the synthesis of new transition metal-olefin and -acetylene complexes continues unabated, only a relatively small amount of data has accumulated on the *thermodynamic stability* of these complexes and these are restricted almost exclusively to complexes of the unsatured species acting as monodentate ligands. Metals able to coordinate strongly with unsaturated ligands are restricted to those in a small triangle around the centre of the periodic table, and designated class (b) acceptors by Ahrland *et al.* [10]. Class (b) acceptors include Cu(I), Rh(II), Ag(I), Pt(II) and Hg(II). However the majority of such metals form inert complexes which are either very readily oxidised or involve solubility problems. If thermodynamic stability constants are to be measured reliably, the equilibrium should be reached reasonably quickly, the reaction should be clean and the stoichiometry should be known or easily deduced. Furthermore, the equilibrium must be followed by means of suitable electrodes or changes in some physical property of the reaction mixture. The solvent is therefore important.

Only for silver(I) complexes is there any detailed information, while for copper(I), platinum(0), platinum(II), palladium(II), rhodium(I) and iron(0) less extensive studies have been made. Although several reviews have been published on transition metal-olefin complexes, emphasis has generally stressed their preparation and not their stabiltiy. The purpose of this section is to survey experimental data in this field and later to correlate them with current theories on bonding. Since the majority of investigations have concerned silver(I) complexes, it is convenient to discuss these first.

2. Silver(I) Complexes

i) Complex Formation in Solution

From a historical point of view, the earliest work followed some studies on the hydration of unsaturated compounds [11] by Lucas *et al.* [12]. They were able to study the following reaction:

$$(\text{olefin})_{CCl_4} + AgNO_{3_{H_2O}} \rightleftharpoons (AgNO_3.\text{olefin})_{H_2O}$$

by comparing the distribution of the unsaturated olefinic compound between aqueous silver nitrate and carbon tetrachloride with the distribution between aqueous potassium nitrate of the same ionic strength and carbon tetrachloride. The reactions were followed by titration. They were able to observe three species, AgL^+, Ag_2L^{2+} and AgL_2^+ where L = olefin, and they believed the stability of the 1 : 1 complex arose from a resonance hybrid structure of the following forms:

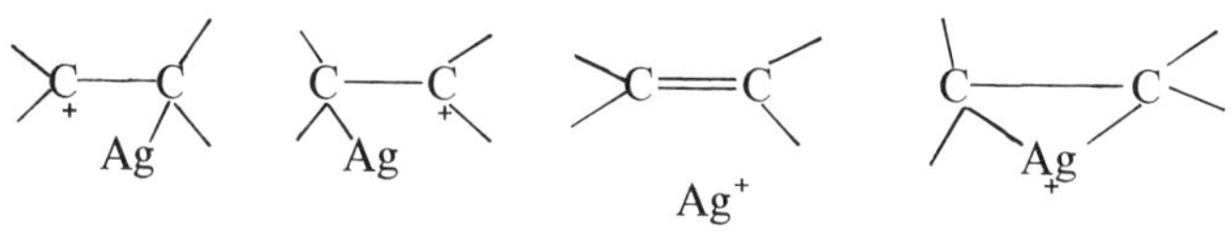

No complex formation was observed with Cd(II), Co(II), Cr(III), Cu(II), Fe(III), Ni(II), Pb(II), Tl(I) or Zn(II).

With silver(I), substitution at the double bond lowered the K value, *e.g.* K is 860 for hex-1-ene, 61.7 for isobutene and 13.3 for trimethylethylene. Conjugated dienes formed less stable complexes than non-conjugated dienes. Studies on mixtures of cis- and trans- pent-2-ene showed the cis isomer to form a more stable complex with silver(I) and a similar study on the 2-butenes again revealed the cis-form to be more strongly coordinated than the trans, although both formed weaker complexes than but-1-ene. The presence of two approximately opposing effects, electronic and steric, was seen from the similar argentation constants for ethylene and propylene; a similar effect was noticed for cis-2-butene and 2-methylpropylene. In general, however, the effect of substituents was thought to be largely steric in nature.

A very approximate value for the heat of silver ion-olefin interaction in solution of ~ -25 kJ mol^{-1} was found for the silver complexes of trimethylethylene and cyclohexene; it was proposed that the silver ion-olefin bond energy approximated to 100 kJ mol^{-1} [13].

The K value for the silver complex of an acetylene, hex-3-yne, as determined by the distribution method [14], was found to be 19.1, *i.e.* smaller than those of alkenes such as the pentenes and cyclohexene, but greater than those of aromatic hydrocarbons [15]. A later study of silver-acetylene complexes [16] using the more rapid solubility technique of Andrews and Keefer [15] gave rise to "quasi thermodynamic equilibrium constants", $K\alpha$ (as opposed to K) for various methyl substituted hex-3-ynes and hept-2-yne. There was good agreement for the K values for hex-3-yne for the two different methods; in each case, replacement of an α-hydrogen atom by a methyl group caused a decrease in the value of $K\alpha$, similar to that observed in alkenes. Values of ΔH approximated to 19–21 kJ mole^{-1}.

By this time, a better picture of the coordination of silver to unsaturated ligands had been proposed by Dewar [17] in his review of the π-complex theory. It was suggested that in addition to the overlap of the filled π-orbital of the olefin with the vacant *s*-orbital or *sp*-hybrid orbital of the silver atom (σ-bond), there was also overlap of the filled *d*-orbitals of silver with the vacant anti-bonding π^*-orbital of the olefin (a dative π-bond). This interpretation of silver-olefin bonding will be discussed more fully in Sect. E. 1.

In some cases, good correlation has been found between formation constants for silver-olefin reactions and reaction parameters for reactions involving similar transition states, *e.g.* heats of hydrogenation of olefins [18]. However, no correlation was found between formation constants for silver-olefin interaction and those for iodine-olefin complexes for a series of cyclic and bicyclic olefins [19]. Interest in this investigation centred on the fact that the formation constants for the silver complexes showed more dependence on *ring size* than did the iodine complexes. Cycloalkenes, methylenecycloalkanes and bicyclic olefins were studied, and the formation constants were essentially in the same order as the estimated relative strains in the olefins. Despite the enthalpies of complexing for the silver-cycloalkene complexes being more negative than those of the silver-methylenecycloalkane complexes, they were less stable due to a more unfavourable entropy change. The majority of dienes appeared to form both 1 : 1 and 1 : 2 olefin-silver ion complexes, although norbornadiene appeared to form only a 1 : 1 complex, which was far more stable than any other olefin studied. This increased stability of the silver complex is due entirely to a more favourable enthalpy change, since the entropy change is even more unfavourable. It is assumed that the silver ion is located under the diene molecule with the resultant overlapping of the appropriate silver ion orbitals with both πelectron clouds.

A previous study [20] on silver complexes of cyclic olefins had revealed that ring strain appeared to facilitate deformation of the π-orbitals for complex formation, *i.e.* the more strained is the cyclic compound, the more reactive it is in electron donating roles. The thermodynamic data for complex formation at

25 °C is shown below in Table 1 and it is interesting to note that the value for ΔH for cyclohexene is very close to that reported by Winstein and Lucas [12a].

Table 1. *Thermodynamic data for complex formation in solution at 25 °C*

Olefin	ΔG (kJ mol^{-1})	ΔH (kJ mol^{-1})	ΔS (J K^{-1} mol^{-1})
Cyclopentene	5.4	-29.4	-117
Cyclohexene	9.9	-23.4	-111
Cycloheptene	9.5	-27.7	-125

From differences in heats of formation, there appears to be a strain of 18.5 kJ mol^{-1} for cyclopentene relative to cyclohexene, and a strain of 17.2 kJ mol^{-1} for cycloheptene relative to cyclohexene.

The most complete study of silver olefin complexes was made by Muhs and Weiss [21], who used a simple, rapid *gas chromatographic technique.* The method made use of the relationship between the complexing equilibrium constants and the retention times for olefins, using a silver nitrate-ethylene glycol stationary phase on the column. The reaction followed was

$$AgNO_{3\,glycol} + \text{olefin} \rightleftharpoons AgNO_3\,(\text{olefin})_{glycol}$$

Three main factors were assumed to be responsible for the variation in stability of silver ion-olefin and -acetylene complexes, namely electronic, steric and structural strain effects but their relative importance was not known, although it is virtually impossible to separate them. Increasing the electron density at the double bond, by exchanging a hydrogen for a methyl group, should increase the stability of the complex, if the olefin is regarded as an electron donor. However, as the opposite is found, the steric influence of substituents would appear to be very important or the π-bond makes a significant contribution.

Muhs and Weiss investigated a large number (121) of aliphatic, acyclic and alicyclic olefins and their main conclusions are assessed below, with the help of Table 2. Increasing alkyl substitution at the double bond reduced the K value, in agreement with the results of Winstein and Lucas [12a] and for any particular homologous series of the various aliphatic olefins, the K value decreased as the carbon number increased. A branched substituent reduced K more than a straight chain substituent, the effect being greater when the group was placed β, rather than α, to the double bond, *i.e.* K decreased in the order

$$R = Me > Et > i\text{-}Pr > t\text{-}Bu,$$

which is the order of increasing bulkiness and increasing electron density at the double bond. The observed order was explained in terms of increased steric hindrance, although if the increased electron density strengthens the σ-bonds while it weakens the π-acceptor bond, then the observed effect could also be explained if π-acceptor bonding was dominant. A similar effect was observed for acetylenes, although substitution of a group in the α-position had little effect; acetylenes gave K values intermediate between corresponding cis- and trans-olefins.

Allenes have a very low affinity for silver ions, while conjugated dienes

Table 2. *Values of K ($l\ mol^{-1}$) for $AgNO_3$ + olefin $\rightleftharpoons$ $AgNO_3$.olefin at 40 °C. K_L = partition coefficient for olefin on pure ethylene glycol column*

Olefin	K	K_L	Olefin	K	K_L
Ethylene	22.3	0.1	1,3-Butadiene	4.2	2.2
Propylene	9.1	0.4	1,4-Pentadiene	10.2	2.9
But-1-ene	7.7	0.9	1,5-Hexadiene	28.8	5.1
3-Methyl-but-1-ene	5.1	1.5	Cyclopentene	7.3	5.8
3,3-Dimethyl-but-1-ene	3.6	2.2	Cyclohexene	3.6	14.7
2,3,3-Trimethyl-but-1-ene	1.8	5.4	Cycloheptene	12.8	27.0
cis-But-2-ene	5.4	1.1	cis-Cyclooctene	14.4	56.1
trans-But-2-ene	1.4	1.0	trans-Cyclooctene	1000	56.1
2-Hexyne	2.0	18.3	2-Norbornene	62	17.7
2-Heptyne	1.6	34.0	2,5-Norbornadiene	33.7	27.8
2-Octyne	1.2	65.2	Benzene	0.1	47.8
Allene	0.8	1.6			

were shown to be poorer complexers than the corresponding mono-olefins, where the weakening of the bonding was thought to be due to delocalisation. K values were much higher for 1,5-dienes, which gave more stable complexes than 1,4-, 1,6-, 1,7- and 1,8-dienes. This may be due to the chelate effect, although potentially chelating diolefins do not always give complexes in which the diolefin is chelating. A high formation constant was again found for 2,5-norbornadiene but not as high as those found for rigid 1,5-dienes. The order of K values for endocyclic olefins is

$$\text{cis-cyclooctene} > \text{cycloheptene} > \text{cyclopentene} > \text{cyclohexene},$$

i.e. different from the found by Traynham and co-workers, suggesting strain energy is not the dominant factor in determining the argentation constants.

Cis-acyclic olefins formed more stable complexes than trans-acyclic olefins and was attributed to relief of strain on complexation; the order of stability is the same as the order of heats of hydrogenation for which a similar explanation was given. The importance of strain was further exemplified when trans-cyclooctene was found to form a much more stable complex than cis-cyclooctene;

the trans-isomer was known to have 39 kJ mole^{-1} extra strain over the cis-isomer.

A further study [22] aimed at obtaining a structure-stability relationship started from the determination of silver formation constants for various substituted styrenes, using both the distribution method and the solubility method. For several *m*- and *p*- substituted styrenes, Fueno and co-workers were able to obtain a reaction constant, $\rho = -0.77$ for a plot of log K v σ, the Hammett function. This indicated an electrophilic reaction and since electron releasing substituents gave rise to more stable complexes, it was suggested that the π-electron density on the vinyl carbons was the most influential factor determining the relative magnitudes of the entropy changes, which in turn controlled the magnitude of log K. The bond order of the external double bond of styrene is intermediate between the bond orders of olefinic and aromatic double bonds. Some thermodynamic data for 1 : 1 complexes is given in Table 3.

Table 3. *Thermodynamic data for silver-ion complex formation in aqueous solution*

	ΔG	ΔH	ΔS
	(kJ mol^{-1})	(kJ mol^{-1})	(J K^{-1} mol^{-1})
p-Methylstyrene	-7.9	-7.9	0
Styrene	-7.2	-8.8	-5.6
p-Chlorostyrene	-5.9	-9.2	-11.0

Little, if any, direct confirmation of the role of *d*-electrons in silver-olefin complexes had been provided, although various theories had drawn conclusions on the relative importance [23] and lack of importance [24] about π-bonding in aromatic hydrocarbons and olefins. The problem was further examined by Fueno *et al.* [25], who concluded that the $p\pi$-delocalisation energy could successfully interpret the observed substituent effects on the silver formation constants of benzene and styrenes. Where the π-bond energy arising from the delocalisation of the silver *d*-electrons was comparable with the σ-bond energy (and with which it varied reciprocally) then the substituent effects were thought to be entropy controlled. Further structure-stability relationships were revealed by their studies on silver-ion complexes of vinyl compounds [26] by the distribution method and of alkenyl alkyl ethers [27] by gas chromatographic means and of cis-stable olefins [28]. Again the inductive effects of polar substituents were the most dominant contributions to the stabilities of these complexes, although steric effects were not necessarily unimportant.

The differences in K values between cis- and trans-complexes were found to be mainly due to differences in ΔH of complexation, in agreement with the results of Cvetanovic *et al.* [29] (see later). For the alkenyl ethers, $R_1R_2C =$ $= CR_3(OR_4)$, the difference in mesomeric effects of the alkoxy groups on complexing was swamped by the inductive and steric effects of the alkyl groups. As substitution of the double bond increased, so the degree of freedom of the coordinated silver ions was correspondingly decreased resulting in a larger loss of entropy, so stability was not directly associated with the strengths of the bonds formed. The relative stabilities of silver(I) complexes of geometrical isomers were found to be dependent upon the relative stabilities of the parent isomer olefins, when the less stable isomer had the greater affinity for the silver ion. *e.g.* 1,2-dibromoethylenes (equal stability), 1,2-dichloroethylenes (trans greater than cis) as in Table 4.

Table 4. *Equilibrium constants for argentation of 1,2-disubstituted ethylenes, $R_1CH{=}CHR_2$ at 25 °C*

Substituent		K		Stable isomer
R_1	R_2	cis	trans	–
Cl	OC_2H_5	0.32	0.59	cis
Cl	Cl	0.25	0.40	cis
Br	Br	0.68	0.56	cis ≃ trans
CH_3	CH_3	62.3	24.6	trans
CH_3	OCH_3	3.13	0.54	trans

Cvetanovic and co-workers [29] first showed that the variation in complex stability was primarily determined by a variation in the enthalpies of complex formation, while the entropies of complexing varied very little. The constancy of ΔS is comparable to that found by Traynham and Olechowski for complexes

Table 5. *Thermodynamic data for silver-ion-olefin complex formation in glycol solutions*

Olefin	$-\Delta G$ (kJ mol^{-1})	$-\Delta H$ (kJ mol^{-1})	$-\Delta S$ (J K^{-1} mol^{-1})
$CH_2{=}CH_2$	7.1	14.6	25.1
$CH_3.CH{=}CH_2$	5.0	14.6	31.4
cis–$CH_3.CH{=}CH.CH_3$	3.9	14.2	34.3
trans–$CH_3CH{=}CH.CH_3$	1.2	10.9	32.2
$(CH_3)_2C{=}CH.CH_3$	0.2	10.0	32.6
$(CH_3)_2.C{=}C(CH_3)_3$	-2.7	7.9	35.6

of cyclic olefins. Table 5 shows a general decrease in $-\Delta H$ values with increasing substitution at the double bond, paralleling the general decrease in stability. The entropy changes oppose complex formation, suggesting that these reactions may be regarded as typical 'soft'-'soft' interactions; the loss of entropy on complexing increased slightly with the increased size of the alkyl substituents, possibly indicating greater physical restraint in complexes containing bulkier substituent groups. Cis-olefins had a greater stability than trans-olefins, due to less negative enthalpy changes for the complexing of the trans-isomers. Comparison was also made of the relative affinities of deuteriated and non-deuteriated olefins for silver-ions; the deuteriated olefins show a somewhat greater affinity due to a more negative ΔH value on complex formation, with the effect being more marked when the deuterium was attached to the α-carbon atom rather than the β-carbon atom. It has been suggested that this may be due to smaller non-bonded repulsions of the deterium atoms than of the hydrogen atoms and the greater inductive effect of a CD_3 group, compared with a CH_3 group.

Temkin and co-workers have investigated the thermodynamic properties of the soluble complexes of unsaturated hydrocarbons with various metal salts with particular reference to their role in *catalytic reactions.* Using a potentiometric technique, they were able to calculate the thermodynamic data shown in Table 6 for the silver(I)-acetylene complexes [30] and the silver(I)-ethylene complex [31]. The results obtained for acetylene have been related to the low activity of silver salts as catalysts for the hydration of acetylene. For the silver(I)-ethylene complex, the relationship between the ionic concentrations and

Table 6. *Thermodynamic data for the reaction*
$Ag^+(aq) + L(aq) \rightleftharpoons AgL^+(aq)$
L = olefin or acetylene

L	$-\Delta G$ (kJ mol^{-1})	$-\Delta H$ (kJ mol^{-1})	$-\Delta S$ (J K^{-1} mol^{-1})	Ref.
C_2H_4	11.3	26.8	52	31)
C_2H_4	11.2	31.8	69	32)
C_2H_4	11.0	–	–	57)
C_2H_2	9.1	55.2	154	

the potential indicated that the complex was of 1 : 1 stoichiometry, but the formation of an $Ag_2(olefin)^{2+}$ species could not be confirmed; the results were also in good agreement with those of Brandt [32] and Trueblood and Lucas [12e]. Free energy and enthalpy data have also been calculated for the interaction of 1,3-butadiene with silver(I) [33] by measurement of the equilibrium constants over a wide range of temperature and ionic strength.

A non-dependence of the thermodynamic equilibrium constant on the solvent for two different types of diols was found [34], which indicated that Ag^+ as well as undissociated $AgNO_3$ formed complexes with olefins, comparable with mercury salt-olefin complexes [35]. Further formation constant investigations [36] by gas chromatography of silver complexes of cyclo-olefins had shown that methyl substitution at the double bond markedly reduced the stability and this decreased further as the substituent became bulkier. In agreement with Traynham and co-workers, cyclopentenes had a higher affinity for silver ions than corresponding cyclohexenes, while for methylenecyclenes, the order of affinity was cyclobutane > cyclopentene ⩾ cyclohexene. Further work [37] on cyclohexene complexes combined with calculations which took into account charge number, ionisation potential, electron affinity, ionic radius, solvation energy and the various atomic excitation energies of the ions demonstrated the selective complex formation of olefins with metal ions having closed *d* shells, *i.e.* Cu^+, Ag^+ and Hg^{2+}. Smith and Ohlson [38] were able to show an approximate parallelism between established complex constants and retention times for hydrocarbons on a silver nitrate-ethylene glycol column; the use of silver nitrate solutions as stationary phases for the gas chromatographic analysis of hydrocarbons has been known for several years. A linear relationship has been found between log K values and the energy difference between the bonding and antibonding π-orbitals for a series of olefins and diolefins; the presence of methyl substituents displaced the bonding orbitals to higher energies, strengthening the σ-bonds but weakening the π-bonds [39].

Several other factors have to be considered in silver-ion olefin complex formation, *e.g.* the variation in solubility of olefins with silver concentration [40]. In dilute solution, the amount of olefin dissolved follows the order $ClO_4^- > BF_4^- \simeq NO_3^-$, whilst in concentrated solution the order is $BF_4^- > ClO_4^- \gg NO_3^-$. The order in concentrated solutions was attributed to protonation of the olefins as a result of hydrolysis of the silver salts, although it may be better related to greater ionic association.

The formation constants for the association of silver ion with the methyl esters of oleic and elaidic acids

(cis- and trans- $CH_3.(CH_2)_7.CH{=}CH.(CH_2)_7.CO_2Me$ respectively)

were determined by the distribution method[41], using isooctane as one phase and aqueous methanol as the other; the cis-isomer was found to be more stable by a factor of 2.5. Potentiometric studies [42] on silver(I) complexes of some unsaturated carboxylic acids, showed that little complexation occurred with vinylacetic, crotonic, fumaric and maleic acids, while for trans-but-2-ene-1,4-dicarboxylic acid, log $K1$ = 3.06 and log $K2$ = 2.49. It was considered that in the latter complex, the oxygens of the carboxyl groups could span the metal to give a linear O–Ag–O arrangement, whereas in vinylacetic acid chelation could not result in a linear arrangement. The introduction of an electron-attracting acetoxy group in the syn-position in 2-norbornene increased the complexation constant, most likely due to chelation with the acetoxy group and the double bond [43]. When the acetoxy group is in the anti-position, the equilibrium constant is considerably reduced, although the retention time in diethylene glycol and in diethylene glycol/silver nitrate columns was greater than that for the syn-isomer indicating that retention times and equilibrium constants are not necessarily easily related.

Table 7. *Silver nitrate formation for the reaction olefin* $(CCl_4) + Ag^+(aq) \rightleftharpoons$ *olefin-*$Ag^+(aq)$

Olefin	K_{aq}
Cyclopentene	0.111
anti-7-Acetoxynorbornene	0.036
syn-7-Acetoxynorbornene	0.400
Norbornene	0.268

An investigation of the silver(I) complexes of a series of acyclic, terminally unsaturated alcohols [44] of general formula, $CH_2 = CH(CH_2)_nOH$, where $n =$ $= 2–6$ and 10, indicated that the equilibrium constant increased to a maximum at $n = 4$ and gradually decreased again to $n = 6$ and 10. Conversion of the hydroxyl group to acetate or methoxy rapidly decreased the stability of the complexes. The enhancement of the stability of the alcohols over the straight chain olefins was attributed to a direct bonding of the hydroxyl oxygen to silver and the geometry of the 1 : 1 silver : 4-pentenol complex was considered to be tetrahedral:

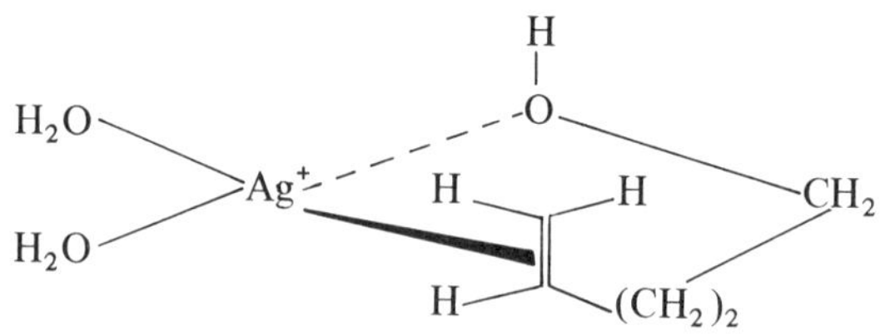

Formation constants for silver(I)-olefin complexes have been obtained in aqueous solution by potentiometric methods by Hartley and Venanzi [45]. The ligands allylammonium perchlorate, but-2-enyl ammonium perchlorate, allyl alcohol and but-2-en-1-ol, were shown to have affinities for silver(I) comparable with acetate and fluoride ions; the complexes are much weaker than the corresponding platinum(II) complexes, possibly due to the non-directional characteristic of the silver component of the σ-bond (5*s*-orbital). The small difference between the formation constants for the allylammonium and but-2-enylammonium complexes and the larger difference between the alcohol complexes was assumed to be an enthalpy effect, arising from an electrostatic repulsion in the case of the unsaturated ammonium cations.

Formation constants of (alkylthio)acetic acids ($RSCH_2CO_2H$) and *p*-(alkylthio)benzoic acids ($RSC_6H_4CO_2H$) with silver(I) have been measured in acetate buffer solutions [46]. By comparison with saturated (alkylthio) acetic acids, the olefinic bond was shown to co-ordinate strongly when the substituent R was but-1-enyl and weakly when the substituent R was pent-1-enyl forming approximately 5- and 6-membered chelate rings respectively. Co-ordination when R was allyl was negligible and, in any case, sterically impossible. It was suggested that the difference in chelation between but-1-enyl and pent-1-enyl was due to steric factors; further investigations [47] have shown the increased stability of the chelate complexes to be due to an enthalpy effect, despite increased opposing entropy changes as chelation becomes more effective.

The relative abilities of some cyclopropyl derivatives of norbornene and norbornadiene to form water soluble silver complexes were studied [48] for the reaction:

$$\text{olefin}_{(\text{pentane})} + \text{Ag}^+_{(\text{H}_2\text{O})} \rightleftharpoons (\text{Ag}^+ \text{ olefin})_{\text{H}_2\text{O}}$$

Norbornadiene formed a more stable complex than norbornene, while corresponding exo- and endo-cyclopropyl derivatives were less stable, due to steric effects.

The solubility of ethylene in aqueous silver nitrate and potassium nitrate solutions has been measured [49] at 30 °C and 0.945 atm. ethylene partial pressure to determine the zero ionic strength association constants for the following reactions

$$\text{ethylene(aq)} + \text{Ag}^+\text{(aq)} \rightleftharpoons \text{ethylene -Ag}^+\text{(aq)} \qquad K_1 = 76$$

$$\text{ethylene (CCl}_4) + \text{Ag}^+\text{(aq)} \rightleftharpoons \text{ethylene -Ag}^+\text{(aq)} \quad K_0 = 2.36$$

K_1 and K_0 are related by K_D by $K_1 = K_D.K_0$ where K_D is the distribution constant for ethylene between carbon tetrachloride and water. The value of K_1, although differing from that determined by Trueblood and Lucas [12e] since a different ionic strength was used, is in close agreement with that determined by Brandt [32] who found K_1 to be 77.2. The value of K_0 as determined by Trueblood and Lucas was 2.44.

ii) Solid Complexes

Relatively fewer studies have been made on solid-phase olefin complexes. Quinn *et al.* [50] investigated the variation of dissociation pressure with temperature of several silver fluoroborate complexes and they were able to calculate the thermodynamic data, listed in Table 8, for the following reaction

$$\text{Ag X (cryst)} + n\text{-olefin (g)} \rightleftharpoons [\text{Ag(olefin)}_n\text{X}] \text{ (cryst)}$$

where X is BF_4^- or NO_3^-.
Ethylene is unique in being able to form stable 1 : 1 and 3 : 2 complexes, since with silver fluoroborate and other monoolefins, 2 : 1 and 3 : 1 com-

plexes are generally preferred. This is presumably due to the small size and symmetry of the ethylene molecule. The value for the heat of formation of the 1 : 1 complex agrees exactly with that found by Tamara *et al.* [51]. Considering the 2 : 1 olefin:silver complexes, it is noted that the heat of formation (ΔH) becomes more negativ as one procedes up the homologous series, despite the entropy changes also becoming less favourable for complex formation. The increased stability from ethylene to butylene is thus opposite to that found in aqueous solution; however, in comparison to solution studies, the cis-2-butene complex is more stable than the trans-2-butene complex, due mainly to the difference in entropy changes. The absence of a 1 : 3 trans-olefin complex has suggested that other steric factors may be involved.

It is interesting to compare the $-\Delta H$ values for the formation of [Ag(1-butene).NO_3] (-42.4 kJ mol^{-1}) obtained by Francis [52] with that obtained by Quinn *et al.* for the corresponding bis-complex, [Ag(1-butene)$_2$.BF_4] (-50.2 kJ mol^{-1}), *i.e.* somewhat lower than half. The decomposition pressure of the solid formed from the interaction of butadiene with silver nitrate indicated the presence of two complexes [53] [Ag(C_4H_6)NO_3] I and [Ag(C_4H_6)$_{0.5}$$NO_3$] II. Their enthalpies of formation were found to be -45.2 kJ mol^{-1} respectively and the proposed structures are as shown below. The more likely structure for II is the *polymeric structure* (a)

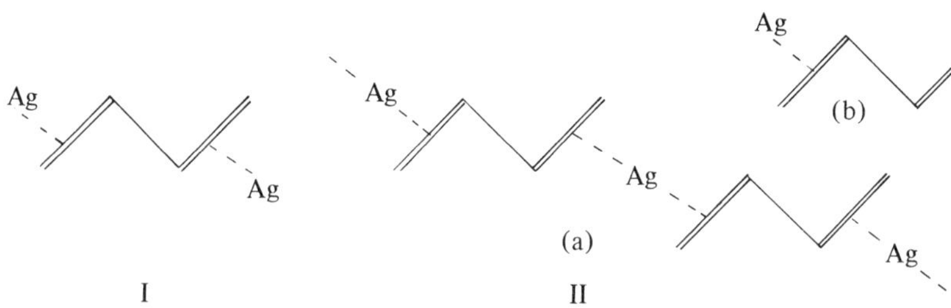

Again, the stability of the complexes were found to vary with the nature of the anion [50] for the propylene complexes.

3. Copper(I) Complexes

Copper (I) complexes of olefins have been less widely studied but have been found to be analogous to silver(I) complexes in several ways. It was shown [54], that solid cuprous chloride absorbed ethylene, propylene and isobutylene and solid cuprous bromide absorbed ethylene to give 1 : 1 complexes, while diolefins (butadiene and isoprene) and acetylenes were reported [56] to form complexes with a 2 : 1 copper:olefin (or acetylene) stoichiometry. Andrews and

Keefer measured the solubility of cuprous chloride in solutions of various unsaturated alcohols [57] and unsaturated carboxylic acids [58], from which they were able to calculate the formation constants for the species $H_2M{\cdot}CuCl$ and $H_2M{\cdot}Cu^{2+}$, although $HM{\cdot}CuCl^-$ and HMCu were sometimes considered (H_2M is a dibasic olefinic acid). By comparison of the solubility of silver bromate in aqueous solutions of unsatured alcohols, it was shown that the cuprous complexes were considerably more stable than the corresponding silver(I) complexes. Whether this is generally true, as has been suggested [59] is open to speculation.

Table 8. *Thermodynamic data for silver-olefin complex formation*

MX(cryst) + n-olefin(g) ⇌ [M(olefin)$_n$X] (cryst)

Complex	$-\Delta G$ (kJ mol^{-1})	$-\Delta H$ (kJ mol^{-1})	$-\Delta S$ (J.K^{-1} mol^{-1})
$Ag(C_2H_4)BF_4$	10.12	44.4	115
$Ag(C_2H_4)_{1.5}BF_4$ (α)	13.72	58.3	149
$Ag(C_2H_4)_{1.5}BF_4$ (β)	13.72	65.1	172
$Ag(C_2H_4)_2BF_4$	15.36	70.8	186
$Ag(C_2H_4)_2BF_4$ (β)	15.36	79.4	215
$Ag(C_2H_4)_3BF_4$	12.89	113.9	339
$Ag(C_3H_6)_2BF_4$	19.20	91.8	244
$Ag(C_3H_6)_3BF_4$	16.23	129.2	393
Ag(but-1-ene)BF_4	21.76	100.6	264
Ag(but-1-ene)$_3BF_4$	21.34	147.9	424
Ag(cis-but-2-ene)$_2BF_4$	25.94	107.9	275
Ag(cis-but-2-ene)$_3BF_4$	26.11	157.6	441
Ag(trans-but-2-ene)$_2BF_4$	17.82	106.8	298
Ag(isobutene)$_2BF_4$	22.09	93.2	238

Substitution of hydrogen at the double bond by methyl groups reduces the stabilty of the complexes formed for both alcohols and acids. However, α, β-unsaturated alcohols showed a greater tendency for complex formation than did α, β-unsaturated acids. It is noted that fumaric acid (where the carboxylate groups are trans) forms a stronger complex with copper(I) than maleic acid (where they are cis); the possibility of coordination through oxygen was also considered.

The first quantitative measurements on copper(I)-ethylene and -actylene complexes were made by Temkin *et al.* [31,60]. The reaction was followed by observing changes in potential of a copper electrode with changes in the equilibrium due to the introduction of the unsaturated hydrocarbon.

$$Cu(powder) + CuSO_4 \rightleftharpoons Cu_2SO_4$$

From a variation in the equilibrium constant for the formation of Cu(ethylene)$^+$ or Cu(acetylene)$^+$, the following data were calculated: Cu(ethylene)$^+$: $\Delta G_{298} = -21.8$ kJ mol^{-1} and $\Delta S_{298} = -50$ J K^{-1} mol^{-1}; Cu(acetylene)$^+$: $\Delta H_{298} = -28.0$ kJ mol^{-1}.

The enthalpy change for the decompositions

$$3\,CuCl{\cdot}C_2H_2 \longrightarrow 3CuCl + C_2H_2 \qquad (1)$$

$$2\,CuCl{\cdot}C_2H_2 \longrightarrow 2\,(3CuCl)\cdot C_2H_2 + C_2H_2 \qquad (2)$$

were calculated from the variation of the vapour pressure against reciprocal temperature [61]. Calculations were also made for the free energy and entropy changes as shown below:

Reaction	ΔG^0 kJ mol^{-1}	ΔH^0 kJ mol^{-1}	ΔS^0 J.K^{-1} mol^{-1}
(1)	1.16	42.7	142
(2)	3.23	42.7	134

For the cuprous chloride-1,5-cyclooctadiene complex, the heat of dissociation was determined as 98.3 kJ mol^{-1} in accordance with the high stability of the complex [62].

The preparation of several cyclodecadiene complexes of both silver(I) and copper(I) has been reported [63]. Whereas the silver(I) complexes could be heated to 100 °C without significant decomposition, the copper(I) complexes began decomposing at 50–60 °C; similarly, the silver(I) complexes could be vacuum dried, whilst the copper(I) complexes decomposed under vacuum or when exposed to conditions of high humidity. The instability of other cuprous chloride and cuprous bromide adducts with cyclic monoolefins and polyolefins had been observed [64], although the relatively stable ones were able to be stored in an inert atmosphere over calcium sulphate. This instability was attributed to the halide ions competing very strongly with the olefins for the copper(I) ligand sites. This supposition was strengthened when Manahan [65] prepared the 1,5-cyclooctadiene complex of copper(I) perchlorate, having a copper: olefin ratio of 1 : 2, by electrolysis at copper electrode in a methanolic solution. For the formation of the complex, log β_1 was evaluated to be 4.5. Perviously [66] he had confirmed the value of log β_1 obtained by Andrews and Keefer for the copper (I) allyl alcohol complex, by a polarographic technique.

Recently a convenient and precise method has been devised for the *determination of thermodynamic data* for copper (I)-olefin complexes [67] based on the coulometric generation of Cu(I) and potentiometric measurement of the Cu(I) activity in a lithium perchlorate-2-propanol medium. The formation constants for the reaction Cu(I) (2-propanol) + olefin (2-propanol) $\rightleftharpoons$ Cu(I) olefin (2-propanol), were found to be linearly related to those of the corresponding

Table 9. *Thermodynamic data for copper(I) cycloolefin complex formation at 30 °C in 1M $LiClO_4$*

Olefin	$-\Delta G$	$-\Delta H$	$-\Delta S$
Cyclopentene	16.6	53.1	120
Cyclohexene	12.1	34.7	75
Cycloheptene	17.5	32.6	50
cis-Cyclooctene	20.1	64.0	145
2-Norbornene	24.7	45.2	68
1,4-Cyclohexadiene	14.7	29.3	48
2,5-Norbornadiene	23.9	62.8	128

silver(I) complexes as determined by Muhs and Weiss [21] although no correlation was found with those as determined by Traynham [19,20]. The thermodynamic data is shown in Table 9. The lack of correlation with Traynham's work was thought to be due to the dependence of the latter's formation constants on the free energy of transfer of the olefin from the CCl_4 to water, while Muhs and Weiss' formation constants were measured in a similar hydroxylic solvent, glycol. The copper(I) complexes were more stable than the silver(I) analogues, generally by a factor of 2–3 log units, in the formation constants. The question of chelation in complexes of 2,5-norbornadiene was also investigated in this study. Comparison of the copper(I) formation constants of 1,4-cyclohexadiene and 2,5-norbornadiene suggested strong chelation with the latter; however this was inconsistent with the results obtained for 2-norbornene. Again, comparison with the results for the silver complex suggested no chelation, reflecting the preference for silver(I) and copper(I) to have linear co-ordination and the inability of 2,5-norbornadiene to occupy two trans sites. This compares favourably with the deduction of Traynham [19,20] although Muhs and Weiss had proposed chelation in the silver(I) complexes. 2,5-norbornadiene can occupy two cis sites, as in the chelated $PdCl_2$ complex [68].

Several complexes of cuprous halides with acrylonitrile [69] and acrolein [70] have been investigated. The enthalpies of complexing have been found from vapour pressure measurements; the enthalpy of formation of the complex from solid copper(I) chloride and liquid acrylonitrile was -29.3 kJ mole^{-1}, while with copper(I) bromide this was -1.3 kJ mole^{-1}. The corresponding value for the acrolein complex was -17.3 kJ mole^{-1}; the enthalpy values for the formation from the gaseous olefinic compounds were -62.3, -34.3 and -49.2 kJ mole^{-1} respectively.

The results of a polarographic and potentiometric study [42] on the affinity of several unsaturated carboxylic acids for various metal ions showed a much greater tendency for bonding through an olefinic linkage for Cu(I) as compared to Cu(II), although in some cases Cu(II) formed more stable complexes than

Ag(I). The same investigation showed no evidence of complex formation between Zn(II) and the unsaturated linkage.

Complex formation between allyl chloride and allyl alcohol and copper(I) has been investigated [71] over a wide range range of temperature and ionic strength, using a potentiometric method which involved the measurement of Cu^{I}/Cu^{0} potentials. It was shown that the complex formation constants of allyl chloride with copper(I) were approximately an order of magnitude lower than the corresponding constants for allyl alcohol and copper(I), suggesting that the catalytic hydrolysis of allyl chloride would show retardation by allyl alcohol in the presence of copper(I). The formation constants for allyl alcohol were found to be independent of the background electrolyte (chloride ions).

4. Platinum Complexes

i) Platinum(II)

Platinum(II) complexes of olefins have been investigated by Orchin and coworkers, when they measured the equlibrium constant for the following reaction, using ultraviolet spectrophotometry [72]:

$$X{-}C_6H_4{\cdot}CH{=}CH_2 + C_{12}H_{24}PtCl_3^- \rightleftharpoons X{-}C_6H_4{\cdot}CH{=}CH_2PtCl_3^- + C_{12}H_{24}$$

$C_{12}H_{24}$ is 1-dodecene

When the equilibrium constants were plotted against the Hammett function for the substituent X, a U-shaped curve resulted, showing all substituents to enhance the stability relative to styrene. This indicated a double bond to present between the metal and olefin, with both the π and σ components contributing equally. Equilibrium constants were of the order 0.03 – 0.05. The relative stabilities of cis- and trans- olefin complexes with platinum(II) were found from a similar study [73], following the reaction

$$(\text{styrene}) + (\text{olefin}) \cdot PtCl_3^- \rightleftharpoons (\text{olefin}) + (\text{styrene})\ PtCl_3^-$$

when the platinum complex of cis-4-methylpent-2-ene was shown to be almost twice as stable as the corresponding trans-complex, and also slightly more stable than unsubstituted cis-pent-2-ene.

Another similar study [74] determined the affinities of 3- and 4- substituted styrenes for platinum(II) by measuring the equlibrium constants for the reaction: (dodecene) + (styrene) $\cdot PtCl_2 \cdot$ 4 – ZPyO $\rightleftharpoons$ (dodecene) $\cdot PtCl_2$ 4 – ZPyO + (styrene) where 4 – ZPyO is a 4-substituted pyridine N-oxide, the substituents being H, CH_3O, CH_3, Cl or NO_2; similar substituents were used in the styrenes. The equilibrium is determined by the substituents on the styrene but to a greater extent by the substituents on the pyridine N-oxide. While electron-attracting substituents on the styrene (*e.g.* $-NO_2$) gave smaller equilibrium constants than electron-donating substituents (*e.g.* $-CH_3O$), the ratio of their equilibrium constants (representing the competition for two styrenes for a site on the Pt) depended on the pyridine N-oxide, *e.g.* for methoxypyridine N-oxide, the ratio for 4-methoxystyrene to 4-nitrostyrene is 210, but only 2 for nitropyridine N-oxide. However, since the equilibrium constants are only accurate to ± 20 %, it is difficult to make generalisations, although it is probable that the σ-bond and π-bond between the Pt and olefin are contributing equally.

A systematic study of platinum(II)-olefin complexes for the equilibria involved in the following reaction was made using a spectroscopic technique:

$$[PtX_4]^{2-} + (\text{olefin}) \rightleftharpoons [PtX_3(\text{olefin})]^- + X^-$$

where X^- = halogen, either Cl^- or Br^-. Several unsaturated amines, substituted amines and alcohols were studied and some of the results are shown in Table 10. Stability constants and enthalpies of formation for platinum(II) olefin complexes of the type

$$(CH_2{=}CH(CH_2)_n \cdot LR_2R')PtCl_3^-$$

L = N, P or As and n = 1 or 2

were obtained [75] in aqueous solution. The olefins had a greater affinity for platinum(II) than water, but lower than amines and comparable with the heavier halides. It was concluded that the π-acceptor of the olefin was more important than the σ-donor capacity for the formation of stable platinum-olefin bonds. By comparison with the corresponding $PtBr_4^{2-}$ system [76], although the stability constants were lower in the case of the heavier halide, analysis of the enthalpy data indicated a stronger platinum-olefin bond for the bromo-complexes. This is most likely due to the fact that the $PtBr_3^-$ group is a poorer σ-acceptor and a better π-donor than the $PtCl_3^-$ group.

Replacement of a hydrogen atom by a methyl group in either the α- or β-position to the double bond caused a reduction in the formation constant [77], the effect being greater in the α-position. Disubstitution further reduced the K values. Analysis of enthalpy and entropy changes indicated both steric and electronic factors to be important in determining the strength of the platinum (II)-olefin bond. Complexes of unsaturated alcohols, which were kinetically the least reactive, gave higher stability constants than corresponding unsaturated amines, and analysis of the enthalpy and entropy changes showed this to be mainly due to solvation effects [78]. However, the anionic allylsulphonate gave complexes of intermediate stability [79], so there appeared to be no regularity of behaviour with regard to charge-charge interactions.

Recently data have become available on the interaction between the cis- and trans-crotylammonium cations and the tetrachloroplatinate anion [80]. The cis-isomer forms the more stable complex, mainly due to a more favourable enthalpy change despite a small increase in the opposing entropy change, although the difference in the thermodynamic parameters for the two isomers does not derive from possible relief of the strain due to bond opposition present in the cis-isomer, as proposed by Muhs and Weiss [21]. The crystal structure of the complexes formed from both isomers are very much similar [80,81].

Table 10. *Thermodynamic data for platinum(II)-olefin complex formation for the reaction* $PtCl_4^{2-} + olefin \rightleftharpoons PtCl_3 \cdot (olefin)^- + Cl^-$

Olefin	Temp (°C)	K	$-\Delta H$ (kJ mol^{-1})	$-\Delta S$ (J.K^{-1} mol^{-1})
all$\cdot NH_3^+$	59.0	1022	29.7	31.8
	44.0	1737		
	30.2	2829		
all$\cdot NH_2Et^+$	59.0	806	24.7	18.4
	44.0	1233		
	24.0	2348		
all$\cdot NHEt_2^+$	59.0	386	23.4	20.9
	45.3	551		
	30.0	865		
all$\cdot NEt_3^+$	59.0	112	20.5	19.2
	45.0	153		
	25.0	260		
but$\cdot NH_3^+$	60.0	2038	21.3	0.8
	44.5	3017		
	30.0	4396		
all$\cdot$OH	60.0	3890	33.9	31.8
	44.5	7250		
	30.0	13011		
all$\cdot SO_3^-$	25.0	4040	25.5	17.2

Ahrland [82] has combined the thermodynamic data obtained by Venanzi *et al.* [75-78] with data given by Elding *et al.* [83] to obtain thermodynamic data for the formation of complexes between trihalogenoplatinate(II) and the soft ligands shown in Table 11. Since the ligands increase in "softness" in the order olefin > Br^- > Cl^-, then the exothermic heat of complexing ΔH, should become more negative in the same order. The enthalpy values for $PtBr_3^-$ are usually more negative, showing that when soft ligands co-ordinate to soft acceptors, the acceptor increases in softness, as discussed by Jørgensen [84].

Using a spectroscopic technique, the equilibrium quotients for the solvation of $Pt(C_2H_4)Cl_3^-$ in water and ethanol were measured at 25 °C and 35 °C [85], when the value for ethanol was 0.02 of the value for the quotient in water in accordance with the weaker solvation in ethanol.

Table 11. *Thermodynamic data for complex formation between trihalogenoplatinate (II) acceptors and soft ligands at 25 °C*

Acceptor		$PtCl_3^-$			$PtBr_3^-$	
Ligand	$-\Delta G$	$-\Delta H$	$-\Delta S$	$-\Delta G$	$-\Delta H$	$-\Delta S$
Cl^-	10.6	18.4	25	14.5	23.8	29
Br^-	17.2	25.5	29	15.8	32.6	59
all·NH_3^+	30.8	48.1	59	–	–	–
all·NH_2Et^+	29.9	43.1	46	29.3	52.7	79
all·NEt_3^+	24.4	38.9	50	–	–	–
all·OH	34.6	52.3	59	–	–	–

ΔG, ΔH in kJ mol^{-1}
ΔS in J. K^{-1} mol^{-1}

ii) Platinum (0)

Chatt, Rowe and Williams [86] studied the displacement reaction

$$Pt(PPh_3)_2\cdot(\text{acetylene}) + (\text{acetylene})' \rightleftharpoons Pt(PPh_3)_2\cdot(\text{acetylene})' + (\text{acetylene})$$

and found the stability of the complexes decreased with the acetylene in the order

$$(p\text{-}NO_2\cdot C_6H_4)\cdot C{\equiv}C\cdot(p\text{-}NO_2\cdot C_6H_4) > PhC{\equiv}CPh > PhC{\equiv}CH > alk\cdot C{\equiv}C\cdot alk > alk\cdot C{\equiv}CH > HC{\equiv}CH.$$

Analogous olefin complexes were unstable. These complexes could be regarded as complexes of platinum(0) or platinum(II), viz:

(a) (b)

The relative stabilities of several bis-triphenylphosphine acetylene-platinum(0) complexes have been determined by kinetic and equilibrium methods on the exchange reaction in cyclohexane using spectroscopic techniques [87]. The rate of both the forward and back reaction is independent of the nature and concentration of the initial free acteylene, suggesting the rate-determining step involves loss of the co-ordinated acetylene. The complexes are stabilised by electron-withdrawing substituents on the acetylene ligand and this, combined with n.m.r. data [88], suggests some degree of double bond character in the metal-acetylene bond, as originally proposed by Chatt *et al.* [89]. The π-acceptor capacity of the acetylene may be more important than the σ-donor ability of the acetylene for the formation of a stable platinum(0)-acetylene bond.

It has recently been shown [90] that the olefin tetracyanoethylene, $(CN)_2C=C(CN)_2$, can displace phenylacetylene from the bis-triphenylacetylene from the bis-triphenylphosphine platinum(0) complex, the first time such a reaction has been achieved. Preparative work on olefin complexes of platinum(0) of the same type has also shown that electron-withdrawing substituents on the olefin increase the stability of the complex, *e.g.* trans-4,4-dinitrostilbene forms a more stable complex than trans-stilbene itself [86].

Metal-ligand interaction in the zero-valent state is stronger than in the divalent state for platinum, the zero-valent metal being a better π-base as expected from the fact that the third ionisation potential is much greater than the first. In zero-valent acetylene complexes, both π-orbitals are involved in bonding and little relative de-stabilisation occurs; however, although the π-bonding and π^*-antibonding levels are energetically closer in acetylene compared to ethylene and would therefore be better σ-donors and π-acceptors so forming more stable complexes, unless they possess bulky substituents that can interact with the metal or the other ligands, acetylene complexes of platinum(II) are not stable and polymerisation of the acetylene tends to occur. In fact, only one π-orbital can act as a Lewis base. In olefin complexes of platinum(0) or (II), only one π-orbital is available (see later).

The possibility of *rotation* of the unsaturated ligand in platinum complexes has been considered recently [91]. The n.m.r. spectra of both Pt(II) and Rh(I) olefin complexes showed rotation of the unsaturated ligand at room temperature, even in the solid state, with a barrier to rotation of $\sim$ 25–60 kJ, indicating that there may be some bond breaking as the process occurs. X-ray studies have shown Pt(II)-olefin and -acetylene complexes to contain the unsaturated ligand approximately at right angles to the plane of the other co-ordinating ligands whereas corresponding Pt(0) complexes contain the unsaturated ligand only 6–14° out of the plane. Calculations have shown that the barrier to rotation is greater for acetylene than for olefin zero-valent complexes, and greater for zero-valent than for divalent complexes. The square planar form was calculated to be the more stable configuration for both zero-valent olefin and acetylene complexes, and n.m.r. spectra indicate that this is maintained even in solution for zero-valent complexes.

5. Palladium(II) Complexes

Solubility techniques have been used for the determination of the thermodynamic values for palladium(II)-olefin complexes [92]; both $PdCl_3 \cdot (olefin)^-$ and $PdCl_2 \cdot H_2O \cdot (olefin)$ were found to be present in solution (analogous with the corresponding platinum(II) complexes) from the following equilibria:

$$PdCl_4^{2-} + \text{olefin} \rightleftharpoons PdCl_3 \cdot (\text{olefin})^- + Cl^- \qquad (3)$$

$$PdCl_4^{2-} + \text{olefin} + H_2O \rightleftharpoons PdCl_2 \cdot H_2O \cdot (\text{olefin}) + 2Cl^- \qquad (4)$$

Some sample experimental data are shown in Table 12. The amount of ethylene absorbed was found to be independent of pH and the enthalpies (and entropies) of complexing were close to zero [92]; the strength of bonding in but-1-ene was similar to that in ethylene and there was no evidence of species such as $(\text{olefin})_2 \cdot PdCl_2$; the values of $K1$ varied slightly with increasing ionic strength, whereas those of $K2$ increased rapidly [92e]. The increase in ΔH as ethylene replaced propylene could arise from steric or electronic effects. Again this could be explained if the π-component is more important than the σ-component of the metal-olefin bond.

Table 12. *Thermodynamic data for the formation of palladium(II)-olefin complexes for the reaction* $PdCl_4^{2-} + \text{olefin} \rightleftharpoons PdCl_3 \cdot (\text{olefin})^- + Cl^-$

Olefin	Temp.	K	$-\Delta H$ (kJ mol^{-1})	$-\Delta S$ (J. K^{-1} mol^{-1})
Ethylene	8.0	15.6	6.3	0
	13.4	16.3		
	20.0	15.2		
	25.0	13.1		
	20.0	16.9		
Propylene	10.3	8.4	0	17
	14.9	8.6		
	20.1	7.9		
	20.0	7.6		
But-1-ene	5.0	13.9	0	21
	10.0	12.6		
	14.8	13.6		
	20.0	12.4		
	20.0	14.3		

Kinetic studies [93] on the oxidation of olefins by aqueous palladium(II) chloride have yielded several equlibrium constants for the reaction:

$$PdCl_4^{2-} + \text{olefin} \rightleftharpoons PdCl_3(\text{olefin})^- + Cl^-$$

The rate expression is

$$-\frac{d[\text{olefin}]}{dt} = \frac{k_1 K1\,[PdCl_4^{2-}]\,[\text{olefin}]}{[Cl^-]^2\,[H^+]}$$

where $K1$ is the equilibrium constant. The order found for $K1$ varied as ethylene > propylene > but-1-ene > cis-but-2-ene > trans-but-2-ene,

(see Table 13) *i.e.* the same order as found for the silver(I) complexes [21], although the differences are quantitatively smaller for palladium(II).

Table 13. *Values for K1 for* $PdCl_4^{2-}$ *+ olefin* $\rightleftharpoons$ $PdCl_3 \cdot (olefin)^- + Cl^-$

Olefin	*K*1	Temperature (°C)
Ethylene	18.7	15
	17.4	25
	9.7	35
Propylene	14.5	25
cis-But-2-ene	8.7	25
trans-But-2-ene	4.5	25
1-Butene	11.2	25

From the present available data, the formation constants for platinum(II)-olefin complexes are approximately 2 log units higher than those for palladium(II)-olefin complexes, although since different olefinic compounds have been used, a true comparison is not possible.

6. Rhodium(I)

The relative stabilities of olefin complexes of rhodium(I) have been compared [94] by following the displacement of ethylene from the thermodynamically stable 2,4-pentanedionatobis(ethylene)rhodium(I), $(acac)Rh(C_2H_4)_2$, using i.r. spectrophotometry:

$$(acac)Rh(C_2H_4)_2 + \text{olefin} \rightleftharpoons (acac)Rh(C_2H_4)(\text{olefin}) + C_2H_4$$

The thermodynamic data shown in Table 14, show replacement of hydrogen by methyl groups in the olefin reduces the stability of the complexes analogous to the trends observed for silver(I) [21,29] and copper(I) [54,55,67] complexes. However, while the 1,2-difluoroethylene complexes are comparable in stability to the ethylene complex, tri- and tetra-fluoroethylene complexes are considerably more stable. Complexes of disubstitued ethylene are less stable if the substituents are on the same carbon atom. Since the effect of an alkyl substituent was about ten times larger for rhodium(I) complexes compared to silver(I), it was proposed that steric hindrance effects were augmented by electrical effects in rhodium(I) complexes; since electron-withdrawing groups strengthened and electron-donating groups weakened the co-ordination of olefins to rhodium(I), the electrical effects were considered to act mainly through the π-bond. This was supported by the observed enthalpy changes. However, the rate of displacement of ethylene by C_2F_4 is $< 10^{-6}$ of the ethylene exchange, suggesting that the formation of the activated complex depends more on the formation of the σ-bond than of the π-bond.

Formation constants for the complexing of ethylene and propylene with tris-triphenylphosphinechlororhodium(I) have been obtained [95]; that for propylene is lower than that for ethylene by a factor of over 10^3.

The affinities of some chelating diolefins for rhodium(I) have been compared by measuring the equilibrium constant of the reaction:

$$\text{RhCl(diolefin)} + \text{diolefin*} \rightleftharpoons \text{RhCl(diolefin*)} + \text{diolefin.}$$

The diolefins studied were
norbornadiene (A),
bicyclo [2,2,2]octa-2,5-diene (B),
2,3,5,6-tetramethyl-7,8-bis(trifluoromethyl)bicyclo[2,2,2]-octa-2,5,7-triene (C),
hexamethylbicyclo[2,2,0]hexa-2,5-diene (D),
COD (E),
COT (F),
2,3-dicarbomethoxybicyclo[2,2,1]heptadiene (G)
and others.

The order of affinities of these diolefins for rhodium(I) is

$$A > B > C \geqslant E > G > D > F.$$

Some of the ratios of the equilibrium constants are $A/B = 3$, $B/C = 2$, $E/G = 80$, $G/D = 100$ and $D/F = 60$ [149].

7. Complexes with Other Metals

A correlation of a high thermal stability with a high ionization potential for a olefin has been observed with iron (0)-olefin complexes of the type $Fe(CO)_4 \cdot$(olefin) [96] *i.e.* poor donor but good acceptor properties increase the stability of the complex. The acrylonitrile complex is one of the most stable, the ethylene complex the least stable and the complexes of styrene or vinyl chloride are of intermediate stability.

The substitution reactions of one acetylene derivative by another in $Co_2(CO)_6C_2RR'$ have been studied [97] and the relative stabilities of the various acetylene derivatives were determined and discussed in terms of the acceptor characteristics of the different acetylenes. R and R' were CF_3, CO_2Me, Ph, CH_2ClMe, $CH_2N(C_2H_5)_2$ and H and in general, the stability gradually increased with increasing electronegativity of R and R' and hence an energy lowering of the acetylene derivative anti-bonding π^* molecular orbitals. The increased stabilisation is therefore linked with a weakening of the σ-bond and a strengthening of the π-bond due to increased back-donation in accordance with Ref. [86] and [89].

Some equilibrium studies have been made on the interactions of mercury(II) with simple olefins. The equlibrium constant for the reaction

$$HgCl_2 + C_2H_4 + H_2O \rightleftharpoons ClHgC_2H_4OH + HCl$$

has been measured [98] by following the uptake of ethylene over a range of temperature. The thermodynamic data are as follows, $\Delta G^0 = 7.2$ kJ mol^{-1}, $\Delta H^0 =$ 15.9 kJ mol^{-1}, $\Delta H^0 = 29$ J.K^{-1} mol^{-1}. The propylene complex was of comparable stability. Kreevoy *et al.* [99] measured the equilibrium constants for the reaction of $HgCl_2$(aq) with cis- and trans-but-2-ene and give values of K_p of 2.62 and 0.9 M^{2}mm^{-1} respectively, while the K_c values are 2.94 and 1.48 respectively. More recently, the formation constant for the mercuric ion-propylene complex has been found to increase with increasing pH [100].

Table 14. *Thermodynamic data for the reaction (at 25 °C)*
$(acac)Rh(C_2H_4)_2$ + olefin $\rightleftharpoons$ $(acac)Rh(C_2H_4)$ (olefin) + C_2H_4

Olefin	Ke	ΔH (kJ mol^{-1})	ΔS (J.K^{-1} mol^{-1})
$CH_2{=}CH{\cdot}CH_3$	0.078	5.9	-2.1
$CH_2{=}CH{\cdot}C_2H_5$	0.092	4.2	-7.1
cis-C_4H_8	0.004	7.5	-20.5
trans-C_4H_8	0.002	7.9	-25.5
iso-C_4H_8	<0.001	16.3	-10.9
$CH_2{=}CHCl$	0.17	3.3	-3.3
$CH_2{=}CHF$	0.32	-6.7	-31.4
trans-CHF=CHF	1.24	–	–
cis-CHF=CHF	1.59	–	–
$CH_2{=}CF_2$	0.10	–	–
$CH_2{=}CHOCH_3$	0.018	–	–
$CHF{=}CF_2$	88	–	–
$CF_2{=}CF_2$	59	–	–
$CH_2{=}CHC_6H_5$	0.08	–	–

A potentiometric method has been used [101] to determine the thermodynamic constants for the interaction of acetylene with mercuric sulphate in H_2SO_4 solution under conditions for the hydration of acetylene. For 1·5–3·OM H_2SO_4 and 0.02 % $HgSO_4$, the data indicated the formation of $HgC_2H_2^{2+}$, for which $\Delta G = -40$ kJ mol^{-1}, $\Delta H = -108$ kJ mol^{-1} and $\Delta S = -230$ JK^{-1} mol^{-1}. Hydration then takes place through the subsequent conversion, which is the limiting step:

$$HgC_2H_2^{2+} \xrightarrow{H_2O} CH_3CHO + Hg^{2+}$$

Baddely has studied the reactions of tetracyanoethylene and other cyano-substituted ethylenes with Ir(I) complexes and found the stabilities of $IrX(CO)(P\phi_3)$ (cyanoölefin) (where X = Cl) from melting points, preparative data and solution behaviour decrease in the order

tetracyanoethylene > fumaronitrile ≫ acrylonitrile ≫ crotonitrile, cinnamonitrile and diphenylmethylene malanonitrile [148].

Only complexes of the first three olefins could be isolated as solids whereas the latter three, possibly due to steric restrictions could not be isolated. The tetracyanoethylene complexes were more stable than the hydrocarbon and fluoro-carbon complexes of Ir(I).

D. Structures of Some Olefin and Acetylene Complexes

1. 'Monodentate' Olefin Complexes

The anion of Zeise's salt has the structure shown in D. I. The dimensions shown were determined in a neutron diffraction study [102] which yielded far more precise results than x-ray crystallographic methods [103]. This is to be expected

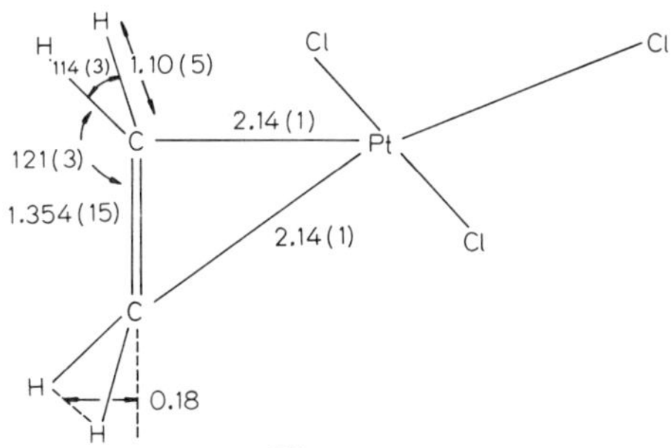

D.I

since the scattering of x-rays by the heavy Pt atom will introduce a large uncertainty into the neighbouring C=C bond length and associated angles. In addition neutron diffraction was able to locate the ethylenic hydrogen atoms. The C=C bond is, within experimental error, at right angles to the $PtCl_3$ plane and may be slightly longer (1.354 ± 0.015 Å) than the equivalent bond in ethylene itself (1.338 Å). The difference, however, is not significant and is in contrast to the marked bond lengthening (to 1.44 ± 0.03 Å) predicted in the x-ray study. As a result of the large uncertainty in C=C distances from x-ray crystallography it is unwise to place any theoretical significance on apparent bond-lengthening in all but the most extreme cases (*i.e.* > 0.1 Å). The carbon atoms in Zeise's salt are equidistant from the platinum atom, the midpoint of the ethylene bond being approximately in the plane of the molecule. Of particular interest is the fact that the four hydrogens are all co-planar and both carbon atoms lie 0.18 Å above this plane, displaced towards the platinum. The hydrogen atoms are therefore bent back significantly showing that the bonding orbitals of the carbon atoms have marked sp^3 character, although standard deviations (±3°) are too large to permit any quantitative estimates. The average bond length for a Pt–C σ bond is approximately 2.05 Å; the Pt–C_2 *pi*-bond has a somewhat greater Pt–C distance (2.14 ± 0.01 Å) although the shortest distance from the platinum atom to the C=C axis is comparable (2.03 Å).

When more than one olefinic bond co-ordinates, as a monodentate ligand, to a metal ion in a planar environment (*i.e. dsp^2* hybridized) they all prefer to be orientated perpendicular to the plane as in Zeise's salt, *e.g.* the dimer $[(C_2H_4)_2RhCl]_2$ appears to have the structure shown in D.II where the planes

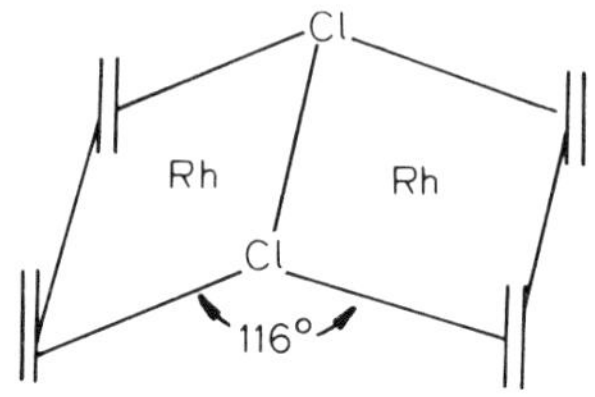

D.II

of the co-ordinated rhodium atoms are inclined at an angle of 116° to one another [104]. This may be due to Rh–Rh bonding. When the olefinic bonds are part of the same molecule, but unconjugated, they co-ordinate perpendicular as far as steric and strain requirements allow, *e.g.* the platinum dipentene complex [105] or the COD analogue (A. II) of the ethylene complex D.II [106]. Norbornadiene is a particularly good example of a ligand with two independent olefinic bonds. The structure of the dichloropalladium(II) complex, determined at low temperature, is shown in D.III [68]. In the silver complex [107] the norbordiene molecules behave as bridging ligands between silver ions as in D.IV.

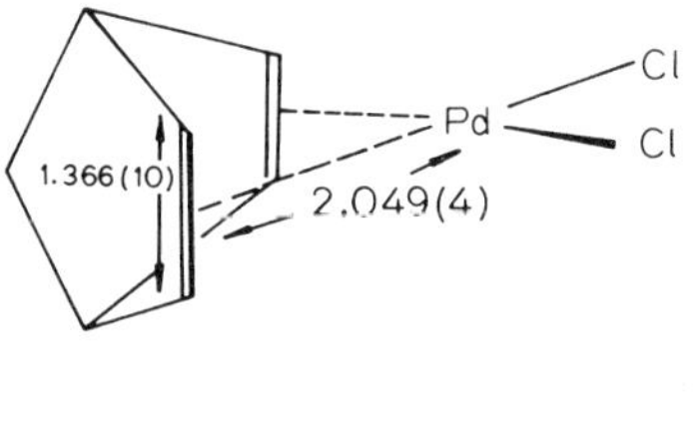

D.III

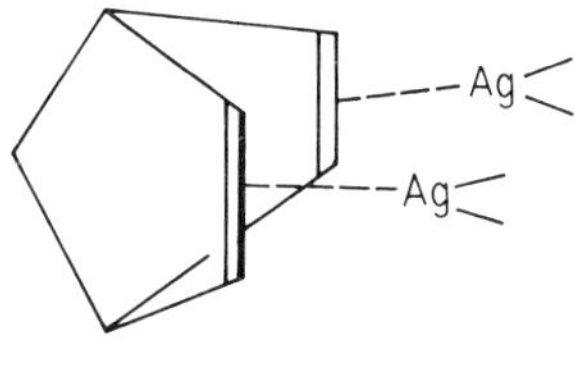

D.IV

A platinum complex in which the ligand contains two independent olefinic bonds, $PtCl_2 \cdot C_{10}H_{14}O_3$, will be referred to in Sect. E.6. Cyclo-octatetraene (COT) may frequently be regarded as a conjugated ring system (*e.g.* in fluxional organometallic molecules [108]) but in many complexes it appears to act as a network of four independent olefinic bonds, co-ordinated to two different metal atoms. For instance, infrared spectral evidence suggests that the polymer [$C_8H_8Rh_2Cl_2$] may have the chain-like structure (D.V) which contains bridging chlorines and a COT ring system [109]. The olefinic bonds in butadiene in the manganese complex shown in D.VI are co-ordinated separately to different manganese atoms and the C(2)–C(3) distance (1.50 Å) is not significantly different from that expected for a normal single bond [110].

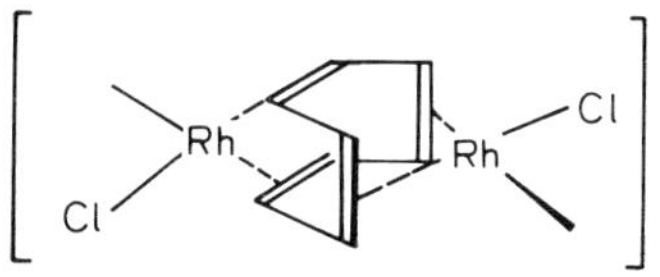

D.V

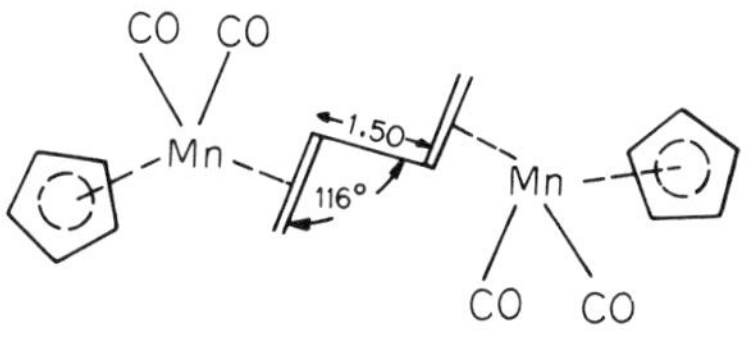

D.VI

Olefins can also behave as monodentate ligands with metal ions hybridized other than dsp^2 (*i.e.* not necessarily planar). Examples include the COD complexes Ni(O) (D.VII) and Cu(I) (D.VIII) [111]. In the Ni(O) complex the nickel

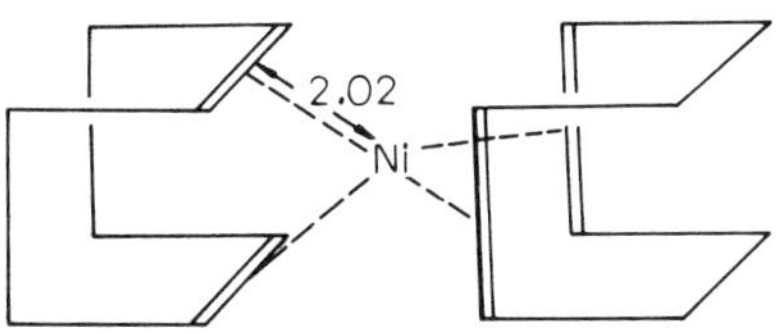

D.VII

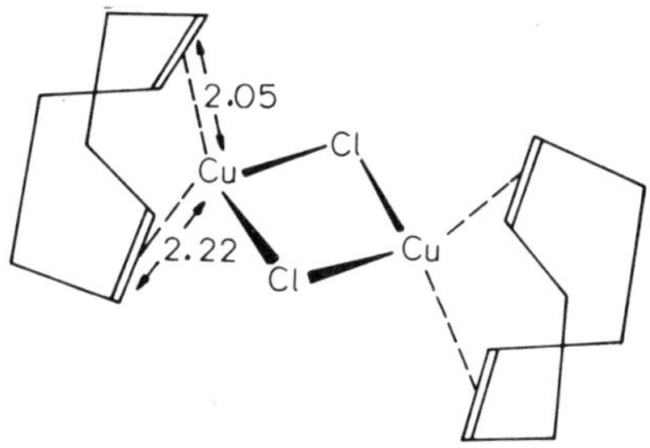

D.VIII

atom lies in a distorted tetrahedral environment. This stereochemistry normally suggests tetrahedral co-ordination hence it is reasonable to deduce that the oxidation state is truely represented as Ni^0 (see Sect. E.3). The platinum analogue, $Pt(COD)_2$, appears to be similar.

Hexamethyl Dewar Benzene (hexamethyl[2,2,0] hexa-2,5-diene, HMDB) can act as a bidentate ligand comparable to COD. The complex of HMDB with $Cr(CO)_4$ (*i.e.* Cr(O)) is shown in D. IX where the two olefinic bonds occupy cis positions of a distorted octahedron [112].

2. 'Bidentate' Olefin Complexes

In this 'bidentate' is used to describe complexes in which a particular olefinic bond appears to occupy two co-ordination positions around a metal ion. In such

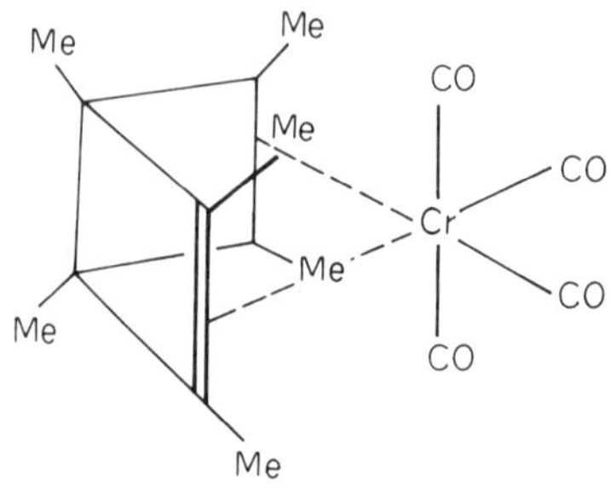

D.IX

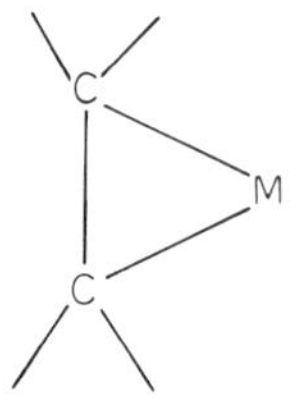

D.X

a case the resulting complex may well be described by the valence bond description shown in D.X in which a pseudo cyclopropane ring containing two C–M σ-bonds is present rather than a *pi*-bond. Reasons for this type of co-ordination will be discussed in section E.3.; as a general rule it is encouraged by the presence of electronegative substituents on the olefin coupled with metals in a low oxidation state.

The formally rhodium(I) complex $[RhCl(C_2F_4)(PPh_3)_2]$ has the structure shown in D.XI [113]. The fluorine atoms are bent back from the metal atom suggesting a change towards sp^3 in hybridization of the carbon atoms (as required in a cyclopropane structure) and the C=C bond is lengthened somewhat. Rh(I) is a d^8 ion so would be expected to favour planar co-ordination. In the complex co-ordination is far from planar being distorted towards tetrahedral, so suggesting an oxidation state other than +1.

Tetracyanoethylene, TCNE, is more clearly 'bidentate' than C_2F_4. Structures of Pt(O) and Ir(I) complexes have both been determined by x-ray methods. The complex $[Pt \cdot C_2(CN)_4 \cdot (PPh_3)_2]$ (D.XII) is virtually planar (suggesting *dsp*^{2}

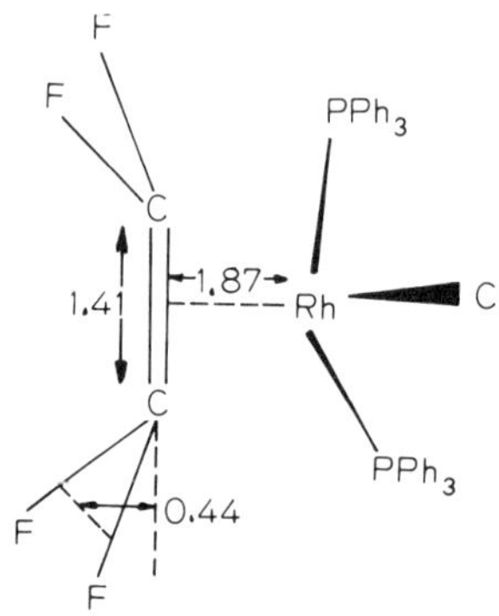

D. XI

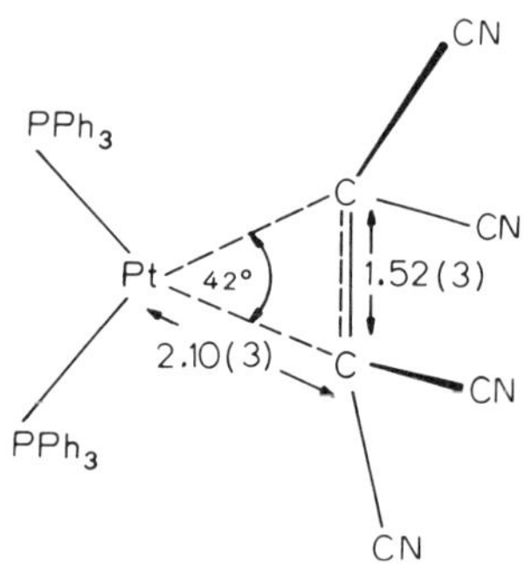

D.XII

hybridization, *i.e.* Pt(II), or sp^2) the C=C bond being tilted 10° with respect to the P–Pt–P plane. The C=C bond is lengthened considerably, being 1.52 Å as opposed to 1.34 Å in TCNE itself. If therefore resembles a C–C single bond [4].

In the complex $[IrBr \cdot CO(C_2(CN)_4)(PPh_3)_2]$ (D.XIII) the C=C bond is also lengthened considerably (1.51 Å) and again the CN groups are bent back considerably [114].

The structure of the Ni(O) complex $Ni(C_2H_4)(PPh_3)_2$ has been determined independently by two groups of workers [115]. Since the reported unit cell dimensions differ slightly the results must be treated with care [116] but both determinations suggest similar structures with the nickel atom in an approximately planar environment, the C=C being tilted 12° to the P–Ni–P plane and the C=C bond being lengthened somewhat (D. XIV). A planar environment for the nickel ion is more compatible with Ni(II) than Ni(O) unless the hybridization is dp^2.

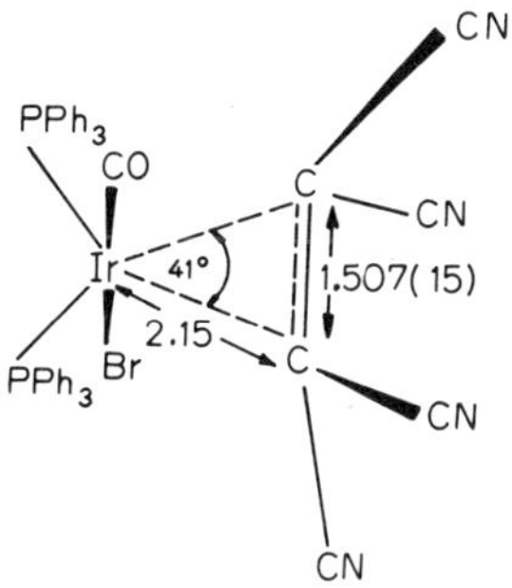

D.XIII

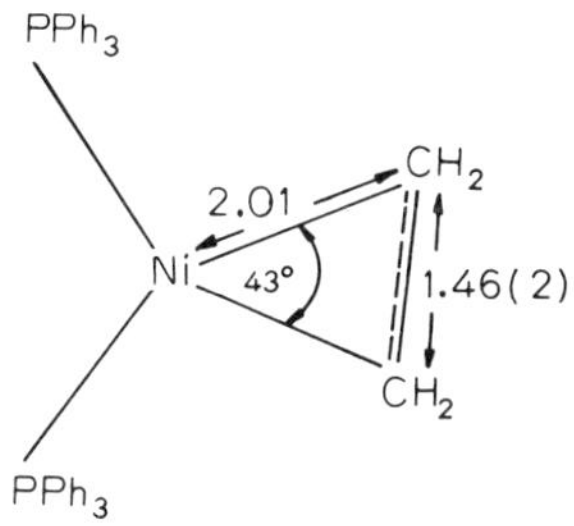

D.XIV

3. Metal-Allene Complexes

In unco-ordinated allenes the C=C=C skeleton is linear, the central carbon atom being *sp* hybridized. On co-ordination the olefinic bonds behave independently and the allene skeleton is bent appreciably. The structure of $[Rh(acac)(H_2C{=}C{=}CMe_2)]$, shown in D.XV, contains two C=C bonds in separate molecules co-ordinated at approximately right angles (103°) to the Rh(acac)

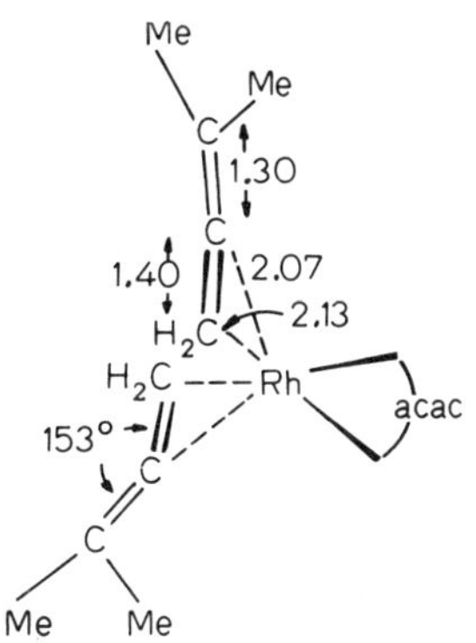

D.XV

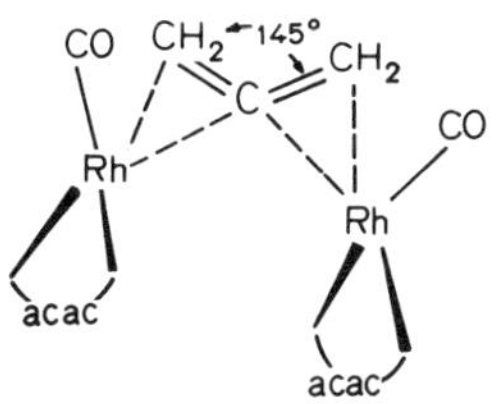

D.XVI

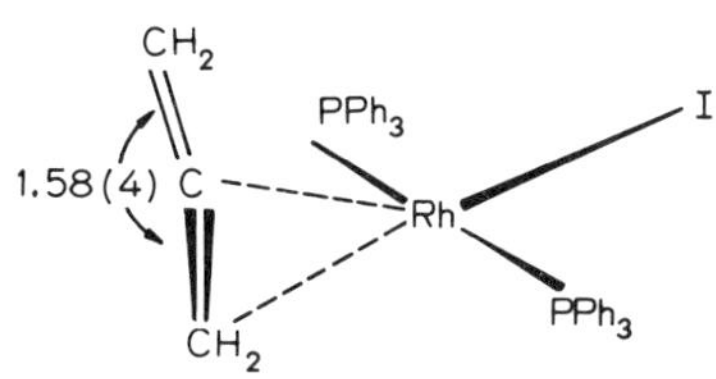

D.XVII

plane. The C=C=C skeleton is bent to an angle of 153° leaving the unco-ordinated C=C bond as a virtually normal double bond [6]. The structure of $[Rh_2(CO)_2(acac)_2(CH_2{=}C{=}CH_2)]$ is shown in D. XVI. Here both C=C bonds of the allene are bonded to different rhodium atoms and the allene skeleton is bent to 144° [6]. Similar geometry is found in the complex $[RhI(PPh_3)_2(CH_2{=}C{=}C{=}CH_2)]$ shown in D.XVII [117]. Allene complexes therefore show a resemblance to those of CS_2 [118].

Allene complexes containing σ-bonded allenes as ligands are also known [119].

4. Metal-Allyl Complexes

While *pi*-allyl complexes are out of the scope of this review many complexes exist in which *pi*-σ-co-ordination is present. In such cases the *pi*-bond resembles a normal metal-olefin *pi*-bond. Examples are the tetramer $[PtCl \cdot allyl]_4$ (D.XVIII) or the dimer $[PtCl \cdot acac]_2$ shown in D.XIX [120].

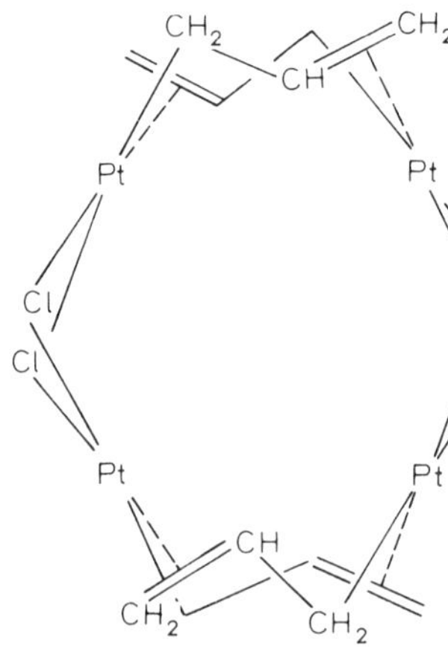

D.XVIII

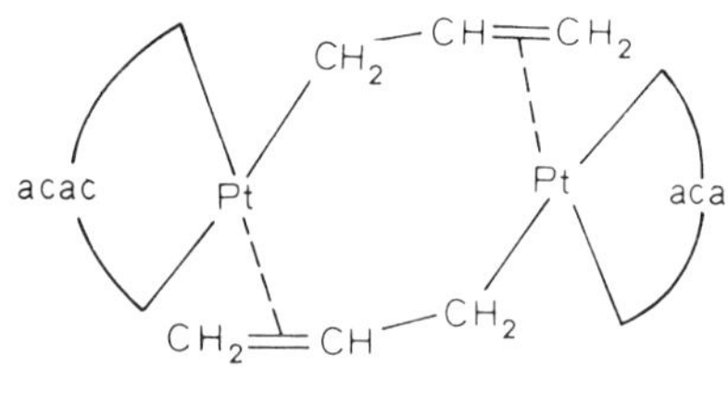

D.XIX

Hexamethyl Dewar Benzene co-ordinates as a *pi*-allylic ligand with palladium (II). The structure of the complex [Pd(acac)dehydro-HMDB] is shown in D.XX [121]. With platinum, however, *pi*-σ-bonding is preferred as in the dimeric planar $[PtCl \cdot dehydro\ HMDB]_2$, D.XXI [122].

It appears to be a general rule that nickel and palladium show a preference for forming normal *pi*-allylic complexes with ligands containing allyl groups while platinum shows a preference for *pi*-σ-bonding of the type shown above. This is possibly a reflection of the increased strength of the Pt–C σ-bond compared to the Ni and Pd–C analogues.

Corresponding *pi*-σ-complexes containing olefinic bonds joined to another donor atom, usually a heavy donor such as phosphorus, are also known. They have recently been reviewed by Nyholm [123].

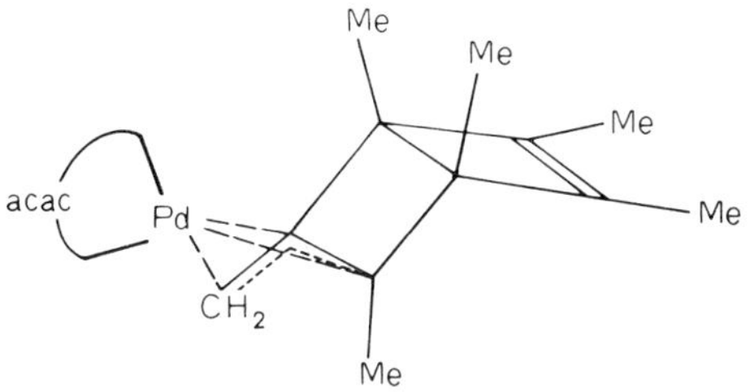

D. XX

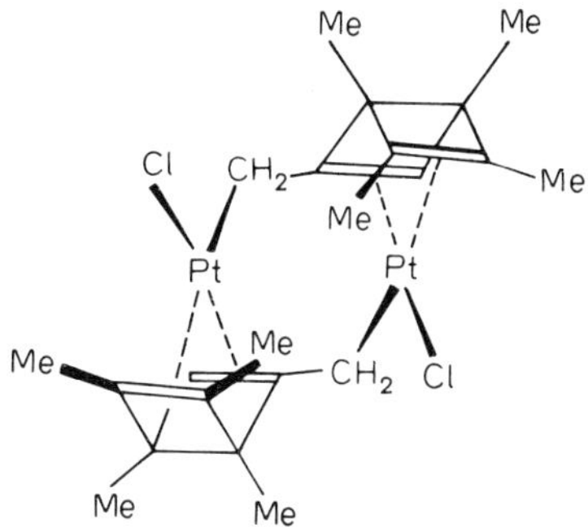

D. XXI

5. Metal-Acetylene Complexes

In the simplest metal-acetylene complexes the orientation of the C≡C bond is similar to that of the C=C bond in Zeise's salt. For example the structure of $[PtCl_2(p\text{-toluidine})(Bu^tC{\equiv}CBu^t)]$ has been determined by x-ray methods is shown in D. XXII [7]. The C≡C bond is perpendicular to the co-ordination plane of the platinum atom and the *t*-butyl groups are bent back from the metal. The C≡C bond length in acetylene is 1.205 Å. In complexes it appears to be lengthened but standard deviations on bond length are generally too great for this bond lengthening to be treated quantitatively.

The complex $[Pt(PPh_3)_2(PhC{\equiv}CPh)]$, a formally Pt(O) complex, resembles the TCNE analogue [4]. It is essentially planar, the angle of tilt being 14°, making the acetylene group appear bidentate. The bond is lengthened considerably (1.32 ± 0.09 Å) and the acetylene molecule is bent to 140° (D.XXIII) [124]. There is therefore a close resemblance between the simpler olefin and acetylene complexes.

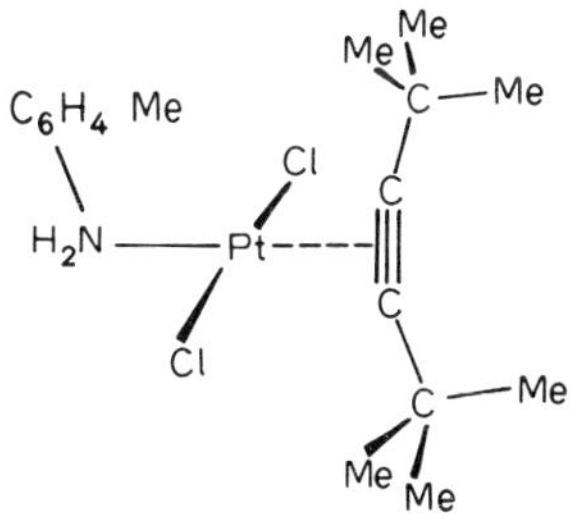

D. XXII

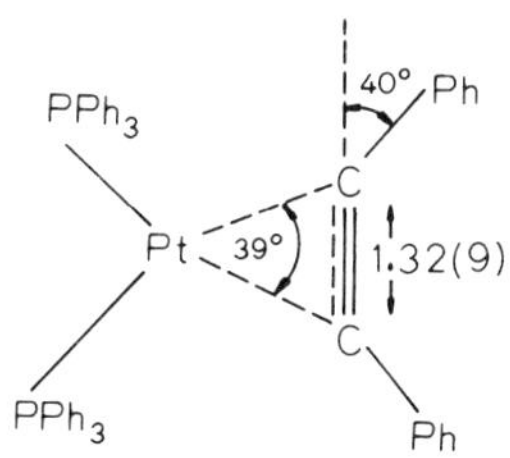

D. XXIII

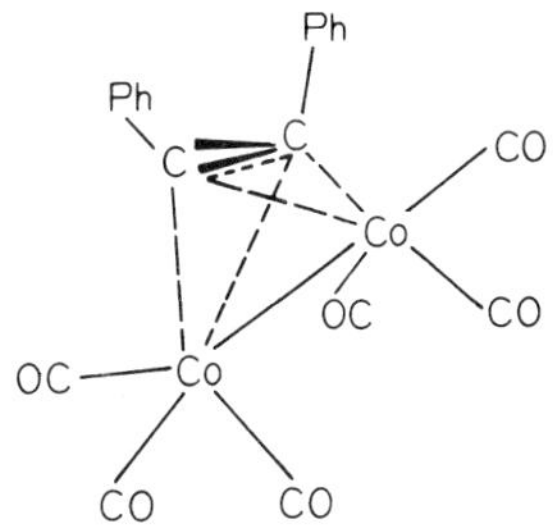

D. XXIV

Since the acetylenic bond can be regarded as involving the overlap of two $p\pi$-orbitals at right angles, it can overlap simultaneously with orbitals from two different metal atoms in planes at right angles to one another. Such overlap is found in [$Co_2(CO)_6$(PhC≡CPh)] shown in D. XXIV [8]. The C≡C bond is now very long (1.46 Å) and the ligand far from linear (138°). The molecule is probably stabilized by a Co–Co bond since the distance (2.47 Å) is less than in $Co_2(CO)_8$ itself (2.52 Å). The C≡C bond is approximately at right angles to this Co–Co bond.

Acetylenes can also form covalent σ-bonds to metals so that mixed *pi*-σ-complexes are possible. A good example of such a compound is [$Co_4(CO)_{10}$(EtC≡CEt)] shown in D.XXV [9]. The acetylene should now be regarded as an olefin with σ-bonds from each of the doubly-bonded carbon atoms to different cobalt atoms and a multicentred *pi*-bond linking the remaining two cobalt atoms and the C=C bond. The C=C bond length is 1.44 Å.

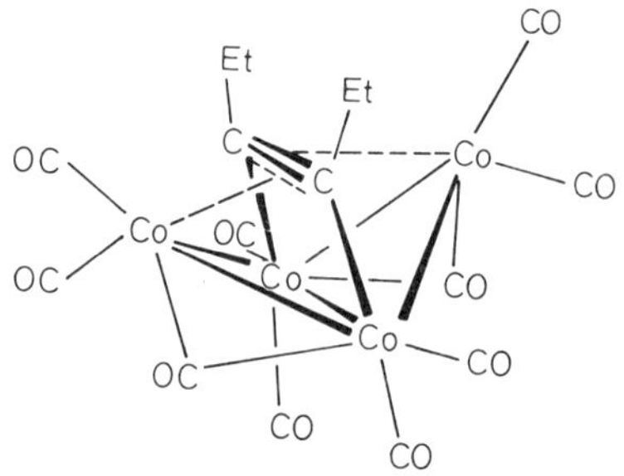

D.XXV

Cyclic acetylenes can also co-ordinate to metal ions. The structure of the complex [$Co_2(CO)_6 \cdot C_6F_6$] is shown in D. XXVI, together with the structure of

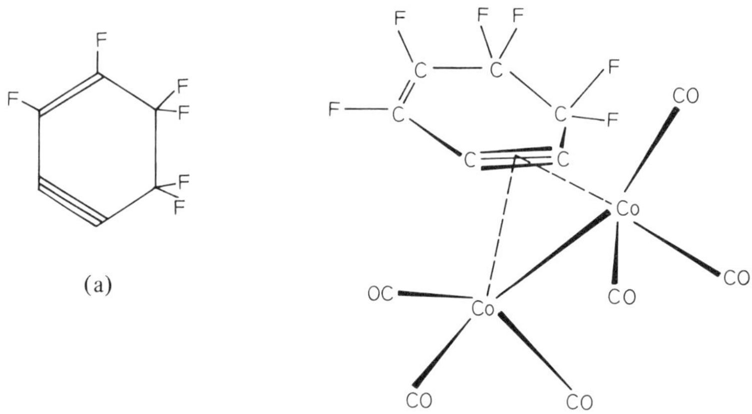

D.XXVI (b)

the ligand itself [125]. The cobalt atoms can be assumed octahedral with a Co–Co bond and the carbon atoms of the formerly triple bond σ-bonded to cobalts leaving a normal C=C bond. This is supported by the measured bond length of 1.36

± 0.03 Å. A true benzyne complex of nickel has been reported (D. XXVII) but as yet no crystal structure determination has been carried out [126].

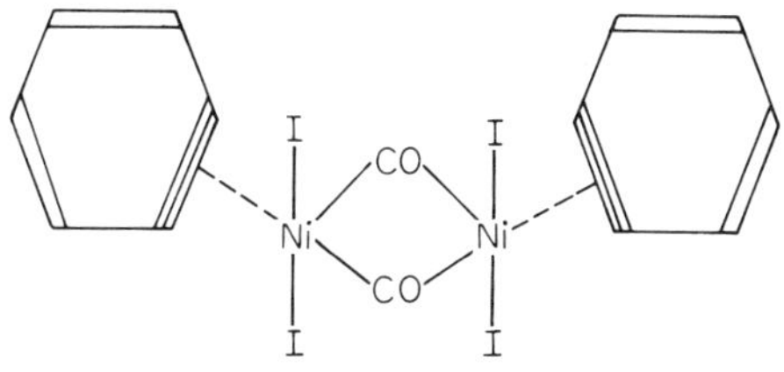

D.XXVII

E. Bonding in Olefin and Acetylene Complexes

1. Silver(I)-Olefin Complexes

The first reasonably satisfactory explanation of the bonding between silver and the C=C bond, based on an application of the MO treatment, was made by Dewar in 1951 as part of a more general discussion of *pi*-complex theory, based predominantly on organic *pi*-complexes [17]. Dewar suggested that the bonding could consist of two components:

i) a normal molecular bond formed by overlap of the filled olefin bonding π-molecular orbital with the empty silver 5*s* orbital and:

ii) an additional bond formed by overlap of a filled metal *d*-orbital (t_{2g}) with the empty antibonding π^*-MO of the olefin as in E.I. The orbitals all have the

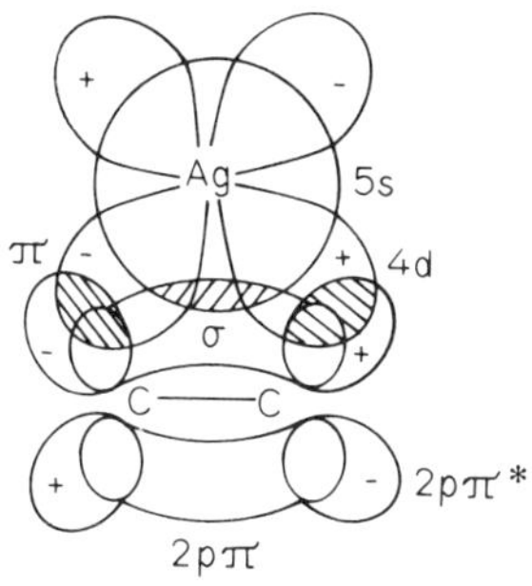

E.I

correct symmetries for such a bonding system of two oppositely directed coordinate bonds. This combination of opposing bonds would leave the olefin much less charged than would be the case in normal organic *pi*-complexes, so accounting for the low reactivity of *pi*-complexes formed from olefins and metals with weakly held *d*-electrons. In addition transfer of electrons into the π^*-orbitals of the olefin will tend to weaken the C=C bond, so tending to cause bond lengthening and a decrease in the stretching frequency, $\nu_{C:C}$. The decrease on co-ordination to silver is 40–70 cm^{-1} [127].

Dewar's theory is still the most acceptable explanation of the bonding in silver-olefin complexes; the relative contributions of the component M←ol (a σ-type bond) and M→ol (a π-bond) co-ordinate bonds is, however, uncertain. Spectroscopic evidence tends to suggest that the σ-bonding is more important than the π-bonding. However the chemical shifts in ^{1}H n.m.r. studies may be explained by either polarization effects or ol→M charge transfer (σ-bond formation). Transfer of 0.1 to 0.2 electrons could account for the entire chemical shifts found but, since polarization effects cannot be calculated accurately, conclusions on the importance of the σ-bonding can be only tentative [128]. Assymetric substitution at the double bond results in a non-symmetric orientation of the Ag^+ ion (from ^{1}H n.m.r.) but the shifts found could result from steric as well as electronic effects [127].

Using ^{13}C n.m.r. more reliable results should be obtained. Results so far suggest that σ-bonding is more important than π although problems in interpretation of chemical shifts are still present [129]. A recent application of e.s.r. to the problem has been interpreted as giving direct evidence of the importance of σ-overlap between the olefin π-orbitals and a vacant *s* orbital on the silver using cycloalkene ligands. The Ag—ol bond appears to decrease in strength as the size of the alkene ring increases [130].

Substituent groups on the olefin with a +I effect (*e.g.* Me) will tend to strengthen the M←ol σ-bond and weaken the π-bond while –I substituents should have the opposite effect. The tendency for the thermodynamic stability of Ag^+—ol complexes to increase as the +I nature of substituents increases supports the claim that σ-bonding is more important than π-back-bonding. However, there is also some evidence that the π-acceptor properties of a ligand are more important than its σ-donor properties [75].

2. Platinum(II) and Other d^8-Olefin Complexes

An MO treatment of the platinum-olefin *pi*-bond was applied by Chatt and Duncanson in 1953 [131]. It is similar to Dewar's explanation of bonding in silver complexes but σ-bond overlap is between a filled ligand π-orbital and an empty $5d6s6p^2$ hybrid orbital of the metal. Back co-ordination from the filled metal $5d6p$ hybrid orbital to the empty olefin π^* orbital assists in restoring a more favourable charge distribution as demonstrated in E.II. A result of the mixing in of a $6p$ with the $5d$ orbital of the metal is that the C=C bond will be perpendicular to the co-ordination plane of the metal atom if overlaps are to be maximised. The orientation of ethylene in Zeise's salt and related compounds

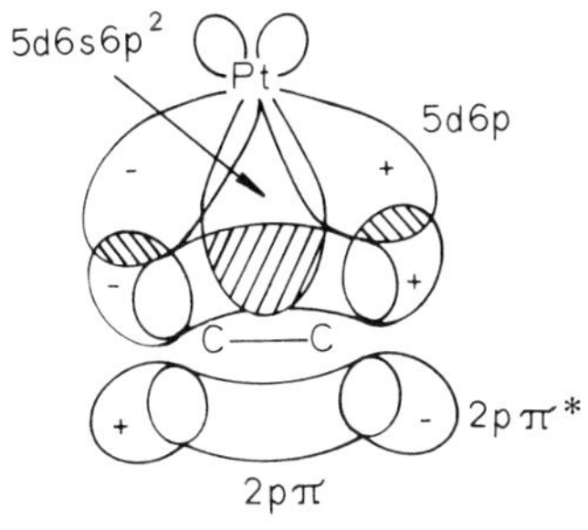

E.II

gives support to this model. Bonding in palladium complexes would be expected to be similar.

As in silver-olefin complexes, charge transfer into anti-bonding ligand orbitals should result in a lengthening of the C=C bond and a decrease in $\nu_{C:C}$. The decrease found in Pt–ol complexes is approximately 140 cm^{-1}; in analogous Pd–ol complexes it is about 120 cm^{-1}. However the use of changes in $\nu_{C:C}$ in interpreting metal-olefin bonding has recently been criticised [132]. Normal co-ordinate analyses of the spectra of some Pd and Pt-olefin complexes have shown the metal-olefin force constants to be low, 2.14 mdyn/Å in $[PdCl_2 \cdot C_2H_4]_2$ and 2.24 mdyn/Å in $[PtCl_2 \cdot C_2H_4]_2$ and $[PtCl_3 \cdot C_2H_4]^-$, and not directly comparable to $\Delta\nu_{C:C}$ [132,133].

A semi-empirical MO calculation has been carried out on Zeise's salt [134]. The orbitals used in the calculation were the *5d, 6s* and *6p* orbitals of platinum, *3s* and *3p* of chlorine and *2p* of carbon. The calculated order of energy levels was found to be reasonable although the calculated transition energies were smaller than the measured values. A Mulliken population analysis of the various orbitals showed a charge of 0.36 electrons in the anti-bonding π^* ethylene orbital. On the other hand σ-bond formation reduced the population of the π-bonding orbital by about 0.33 electrons giving the net effect of keeping the ethylene nearly neutral and suggesting that the π- and σ-bonds contribute approximately equally.

Iron(0) is isoelectronic with Pt(II). Infrared and mass spectra of the complexes $[(olefin)Fe(CO)_4]$ have been correlated to give a ratio of the donor-acceptor strengths of a range of olefins [96]. This is discussed in Sect. C.7.

A ^{13}C n.m.r. study of a number of rhodium-olefin complexes has recently been reported [135]. The magnitude of the ^{103}Rh- ^{13}C coupling constant and the upfield shift of the olefinic carbon resonance on complexation have been interpreted as suggesting substantial *s*-character (σ-bonding) in the olefin-rhodium bonds – possibly of the order 15 %. If the hybridization of the metal is thought to be dsp^2 this implies that σ-bonding contributes 60 % to the total bonding. A similar study of the complexes $[(Ph_3P)_2Pt(C_2H_4)]$ and $[(Ph_3P)_2Pt(C_2H_2)]$ suggests, by similar arguments, contributions of up to 100 % form the σ-bonded scheme, making π-bonding insignificant [136].

Very recently semi-empirical MO calculations have been carried out on olefin and acetylene complexes of Pt(O) and Pt(II) [137,138]. These will be discussed more fully in Sect. E.3. However the result of calculations on the mole-

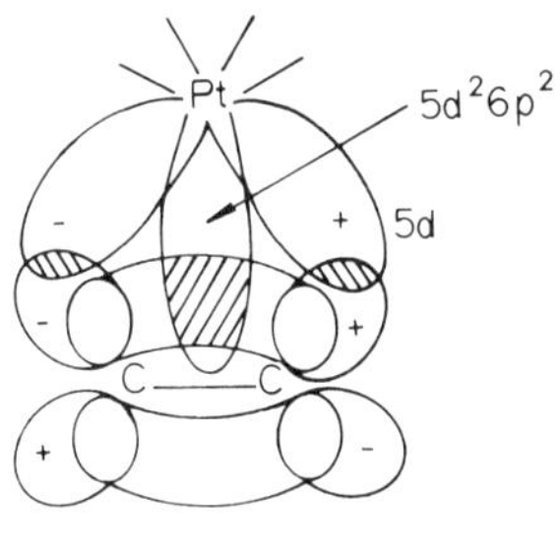

E.III

cule $[(NH_3)Cl_2Pt(C_2H_4)]$ suggest that the hybridization of the valence orbitals of the Pt(II) should be d^2p^2 and not dsp^2, both of which would give a square planar environment. Orbitals participations in the bonding are shown in Sect. E.III. The d_{xy}, $d_{x^2-y^2}$, p_x and p_y orbitals form four σ-bonds while orbitals in the xz and yx planes participate in π-bonding. When the dihedral angle (*i.e.* the angle between the co-ordination plane of the metal and the C : C bond) is 90° this π-interaction is maximised, stabilising the pseudo-tetrahedral configuration found in Zeise's salt. Calculations were carried out for angles of 0, 22.5, 45, 67.5 and 90° to give a potential energy curve with minima at 0° and 90° that at 90° being more favourable as shown in Sect. E.Va. The barrier to rotation has been found experimentally to be 40–60 kJ mol^{-1} [139,140]. The calculated values is in reasonable agreement (approx. 100 kJ mol^{-1} [138]). Rotation of co-ordinated olefins and acetylenes will be discussed in Sec. E.5.

3. Platinum(0) and Other d^{10}-Olefin Complexes

The x-ray crystal structures of zerovalent olefin complexes indicate an approximately square-planar co-ordination [4,114,115] with a dihedral angle of less than 14°. Such $[ML_2(\text{olefin})]$ complexes can then be thought of as either three co-ordinate zerovalent complexes or four co-ordinate divalent complexes containing two M–C σ-bonds (D.X). The structure of the complex $[Pt(PPh_3)_2TCNE]$ (D.XII) supports the latter, since the C–C bond length is normal for a C–C bond and the carbon atoms appear to approach sp^3 hybridization. The complex may therefore be thought of as involving a $d{\rightarrow}\pi^*$ back-bond with virtually no TCNE→Pt σ-bonding. Such complete Pt→TCNE charge transfer into the ligand π^*-orbital would make the C–C bond a single bond as found. TCNE is one of the strongest '*pi*-acids' known, hence would be expected to encourage back co-ordination. Mathematically an MO description in terms of a pure π-bond and the VB description in terms of two σ-bonds are the same but chemically the C–Pt–C bond angle of only 42° is not compatible with the high thermal stability of the complex. Similar arguments would apply to the iridium(I) analogue [115].

The ligand C_2F_4 would be expected to show some of the properties of a strong '*pi*-acid'. This appears to be the case from the structure of the complex $[RhCl(PPh_3)_2(C_2F_4)]$ (D.XI).

Bonding in these zerovalent complexes has been explained by the semi-empirical MO calculations of Nelson *et al.* [137,138]. The results of these calculations indicate that the Dewar-Chatt-Duncanson model can be extended to cover both zerovalent olefin and acetylene complexes with small modifications to the hybridization schemes involved. The two extreme types of co-ordination referred to earlier, *i.e.* three co-ordinate zerovalent or four co-ordinate divalent, are synthesised into a consistent explanation based on dp^2 hybridization of the metal ion, giving planar trigonal hybridization by using the d_{xy}, p_x and p_y orbitals. The metal ion should therefore be thought of as M(0) with the olefin occupying only one co-ordination site. Calculations of the total energy of the complex as a function of the dihedral angle are in agreement with the planar configuration found (see Sect. E.5). Orbitals participating in the bonding will be similar to those illustrating the Pt(0)-acetylene bonding (E.IV).

The semi-empirical MO calculations also gave an indication of the ratio of σ-bonding: π-back-bonding. Both Löwdin and Mulliken population analyses were carried out to give electron densities on the platinum atoms. Back-bonding would tend to leave the metal positively charged while σ-donation would have the opposite effect. Calculated charge densities for four different ligands [138] are:

Molecule	Charge on Pt	
	Mulliken	Löwdin
$(PH_3)_2Pt(CH_3\cdot C\vdots C\cdot CH_3)$	+1.58	-0.08
$(PH_3)_2Pt(CH_3C\vdots CH)$	+1.54	-0.07
$(PH_3)_2Pt(CH_2:CH_2)$	+1.59	-0.04
$(PH_3)_2Pt(TCNE)$	+2.17	+0.09

Values obtained by the two methods are very different, Nelson *et al.* favour the Löwdin figures [138], but the important fact is that the trends are the same in each case and are in agreement with the trend expected from the electronic effects of the substituents, Pt→TCNE showing more back donation than Pt→C_2H_4. Recent ^{13}C studies on Pt(O) complexes have been interpreted as giving information on the *s*-character of the Pt–C σ-bonds [136]. However the results are difficult to interpret, largely as a result of the inadequacy of the valence bond approach to bonding in these complexes. The ^{13}C studies have, however, allowed calculations of the carbon bond angles which are in reasonable agreement with measured angles (see Sect. E.6).

4. Metal-Acetylene Complexes

Bonding between acetylene and metals appears to be essentially similar to metal-olefin bonding. However acetylene has two π-orbitals in planes at right angles to one another and so can *pi*-bond simultaneously to two metal atoms. The chemistry and early theories of bonding of metal-acetylene complexes has been reviewed by Greaves, Lock and Maitlis [3] and the bonding theory of Nelson, Wheelock, Cusachs and Jonassen described in Sect. E.3 applies equally well to acetylene complexes [137]. Assuming dp^2 hybridization of the metal atom the bonding will be that shown in E.IV. Calculations of the energy required to ro-

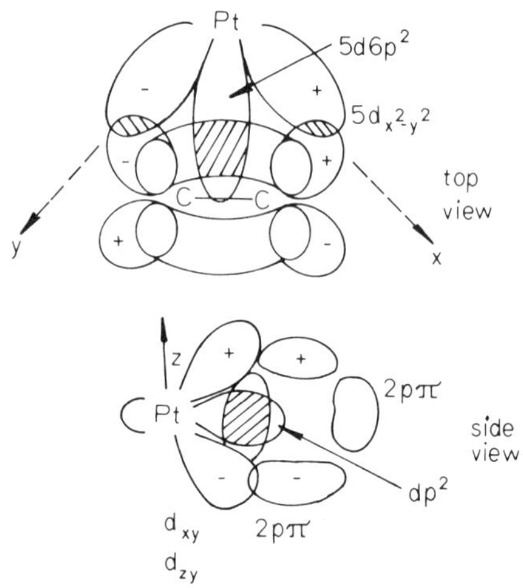

E.IV

tate the acetylene show no minimum at a dihedral angle of 90° and predict the planar co-ordination found in all zerovalent acetylene complexes. The barrier to rotation appears to be greater than in the Pt(II) complexes [138]. By using the hybridization schemes of dp^2 for Pt(0) and d^2p^2 for Pt(II) the tendency for Pt(II) acetylene complexes to undergo polymerisation may be explained. In the Pt(0) complex both acetylene π-orbitals are, to some extent, involved in bonding to the metal while only one is involved with Pt(II), so destabilizing the other by a synergic mechanism and allowing the complex to act as a Lewis base.

5. Rotation of Co-ordinated Olefins and Acetylenes

Olefin rotation was first observed in some ethylene-rhodium(I) complexes using ^1H n.m.r. spectroscopy [139]. At room temperature the complex $[Rh(Cp)(C_2H_4)_2]$ shows two broad peaks at 8.7 and 6.9 τ which are due to pairs of non-equivalent protons undergoing rotation at an intermediate rate on the n.m.r. time scale. This rotation can be frozen out at low temperatures and accelerated at higher temperatures to give a calculated activation energy of 62.7 ± 0.8 kJ mol^{-1}. When one of the co-ordinated ethylenes is replaced by C_2F_4 or SO_2 the barrier to rotation is reduced somewhat to 56.8 ± 2.4 and 51.0 ± 3.2 kJ mol^{-1} respectively. There was no evidence of rotation of the C_2F_4 ligand below the decomposition point of 100°.

Olefin rotation has also been observed in platinum(II)-olefin complexes. ^1H n.m.r. studies on Zeise's salt suggest some motion of the ethylene [141] and recent n.m.r. studies on the complexes [Pt(acac)Cl(olefin)] for a range of olefins have allowed the calculation of activation energies which fall within the range of 45 – 65 kJ mol^{-1} [140]. The preferred orientation of the olefin is at right angles to the co-ordination plane of the platinum, as expected from the structure of Zeise's salt. Activation energies for rotation of the olefins were found to lie in the order

trans-but-2-ene > propylene ~ cis-but-2-ene ~ ethylene > tetramethylethylene.

This is close to the order expected to result from changes in steric interaction during rotation since the tetramethylethylene ligand has no position of great steric stability.

Rotation in solutions of zerovalent complexes is not so well established. ^1H n.m.r. spectra of complexes of unsymmetrical acetylenes and phosphines show coupling to two nonequivalent phosphorus atoms showing that the square planar environment is maintained in solution [3]. However rotation of both ethylene and acetylene in zerovalent complexes has been predicted [115].

Rotation in both zerovalent and divalent platinum complexes has been considered theoretically and the calculated barrier to rotation was found to be greater for acetylene than olefin zerovalent complexes and greater for zerovalent than divalent complexes [138]. For the divalent complex [trans-$(NH_3)PtCl_2(C_2H_4)$], two energy minima were predicted with dihedral angles of 0° and 90°, that at 90° being lower. This is illustrated in E.V(a). The existence of a second energy minimum when the ligand is in the co-ordination plane (*i.e.* when the angle is 0°) suggests some bond breaking in the process of rotation and reduces the size of the activation energy to the reasonably accessible value of 1.1 eV (about 100 kJ $mole^{-1}$). The total energy of the zerovalent complexes

$$[(PH_3)_2Pt(CH_3C{:}CCH_3)] \text{ and } [(PH_3)_2Pt(CH_2{:}CH_2]$$

as a function of the dihedral angle show only one energy minimum at 0° in accordance with experimental observations. This is illustrated in E.V(b), (c). The activation energy is now considerably larger, being 330 and 310 kJ mol^{-1} respectively. The absence of a minimum at 90° has been interpreted as the result of a continuous rehybridization of the orbitals involved in π-bonding upon rotation rather than bond breaking. The higher barrier to rotation in the acetylene complex (b) compared to the ethylene analogue (c) supports the claim that π-bonding is more important in acetylene *pi*-complexes. Reactivity of co-ordinated olefins is obviously connected with ease of rotation. This will be referred to in Sect. E.7.

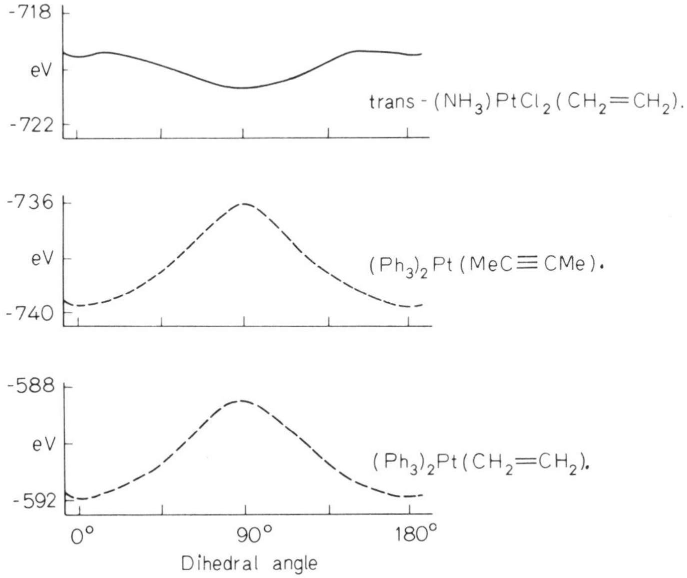

E.V

6. Geometry of Co-ordinated Olefins and Acetylenes

Ethylene itself, in its ground state, is planar while acetylene is linear. The geometry of both molecules is changed on co-ordination. Neutron diffraction studies on Zeise's salt show the hydrogen atoms to be co-planar and bent away from the metal atom [102] and similar distortions are to be found in all structures of co-ordinated olefins and acetylenes reported.

Distortion of this type is to be expected as a result of normal steric interaction, particularly when bulky groups are attached to the olefinic bond. However in complexes such as $[Pt(PPh_3)_2TCNE]$ (D.XII) the distortion from planarity is very large indeed. A VB interpretation of such distortions would describe them as the result of a partial rehybridization of the orbitals of the carbon atoms from sp^2 to sp^3 for olefin complexes and sp to sp^2 for acetylenes. In some molecules (*e.g.* $[Co_4(CO)_{10}(EtC{:}CEt)]$ (D. XXV)) this is a reasonably acceptable explanation but it is not so helpful when distortions are small.

A more satisfactory explanation was presented to account for the structure of $[Pt(PPh_3)_2CS_2]$ which contains a co-ordinated S=C bond as illustrated in E.VI [118]. While not an olefin itself, co-ordinated CS_2 closely resembles co-ordinated allenes (see Sect. D.3) and the arguments can be extended to cover simpler olefins and acetylenes. Points of interest in the complex (E.VI) are:

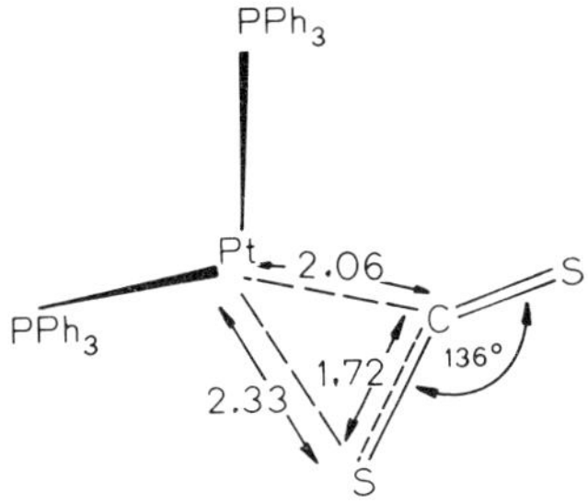

E.VI

(i) The S–C–S angle of 136° is almost identical with that found, by spectroscopic methods, for the first excited state of CS_2 and

(ii) the co-ordinated CS bond length (1.72 Å) resembles the bond length in the first excited state of CS_2 (1.64 Å) while the unco-ordinated CS bond (1.54 Å) is the same as that of CS_2 in the ground state (1.55 Å). The Pt–CS *pi*-bond can therefore be thought of as involving a one electron transfer from the highest π-MO of the ligand to its lowest π^*-MO, giving the ligand the geometry of the $^3A_2$ state – the first excited state. The observed bond angles in the C–C–C skeleton of co-ordinated allenes [6] are in general agreement with the above.

The geometries of complexes of unsaturated ligands have been examined by means of a simple MO treatment by McWeeney, Mason, and Towl [142,143]. They concluded that the charge distribution should be close to that possessed by the ligand in its first excited triplet state. Hence on co-ordination the geometry should change spontaneously to approach that of the triplet state rather than the ground state. For free acetylene the first excited state would be trans-bent as a result of non-bonded interactions while co-ordinated acetylene would obviously prefer a cis-bent structure. It has been estimated that the bond angle in the cis-bent excited state would be 142° and the bond length 1.38 Å [144]. This is in good agreement with the measured angle of 140° in $[Pt(PPh_3)_2PhC{:}CPh]$ [124]. Simple MO calculations indicate that the bond angles found in co-ordinated acetylene can be accounted for by the transfer of 0.5 electrons from the ligand π_u to the π^* orbitals.

A VB description in terms of a partial change in hybridization of the multiply bonded carbon atoms can, in some ways, be regarded as comparable to the MO description in terms of excited states *e.g.* sp^3 hybridization would be an excited state for the carbon atoms in ethylene. The VB description has been introduced into interpretation of ^{13}C n.m.r. spectra of metal-olefin complexes [135] and has permitted the calculations of some bond angles of co-ordinated ligands [136]. The C–C–H angle in co-ordinated acetylene was calculated to be 139° in the complex $[(PPh_3)_2Pt{\cdot}C_2H_2]$ in excellent agreement with comparable experimental values, and 115° in the corresponding ethylene complex. The measured angle is 121 ± 3° [102].

Since the allene group can co-ordinate through just one of the two double bonds, unsymmetrical allenes (*e.g.* $Me_2C{:}C{:}CH_2$) could give two different products on co-ordination. The structure of the complex $[Rh(acac)(CH_2{:}C{:}CMe_2)_2]$ (D.XV) shows that bonding is to the olefinic bond furthest removed from the methyl groups. In the complex $[PtCl_2(C_{10}H_{14}O_3)]$ (E.VII) the Pt–C bond lengths

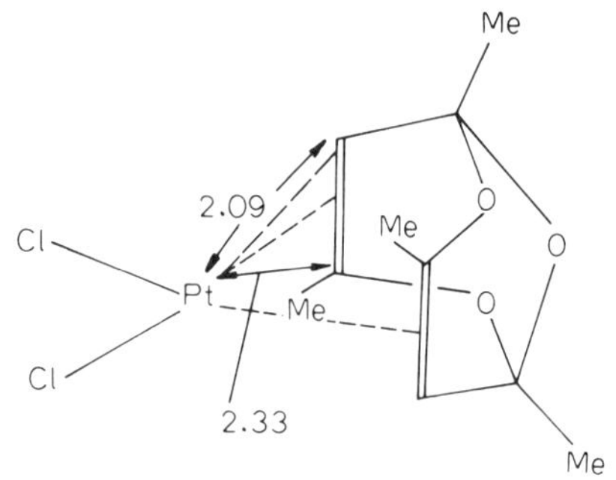

E.VII

to the carbon atoms bonded to methyl groups are longer (2.33 Å) than those to the unsubstituted carbons (2.09 Å) [145]. Both of these facts suggest that methyl groups destabilize metal-olefin bonds. Since the methyl group has a + I inductive effect it will tend to destabilize the π^*-orbital of the olefin so decreasing the overlap between the filled metal *d*-orbitals and these ligand antibonding orbitals. This destabilization is added experimental evidence of the importance of back co-ordination in the metal-olefin bonds.

7. Reactions of Co-ordinated Olefins and Acetylenes

Reactions of co-ordinated olefins and acetylenes are very different from those of the free ligands. Classes of reaction have been summarised by Hartley [1]. The reactivity of these co-ordinated ligands has been explained by Mano and others as resulting from the *d*-orbitals of the metal atom causing a shift of the reaction from a 'forbidden' to an 'allowed' category. For example, a correlation diagram can be drawn up which accounts for the conversion of 2 complexed ethylene molecules to cyclobutane [146]. The role of the metal atom in these ligand reactions may be visualized as destroying the π-symmetry of the olefin orbitals by co-ordination. On a VB description there will then be a partial rehybridization from sp^2 to sp^3 which will weaken and hence lengthen, the double bond. The ligand is now in a distorted or excited state and its orbitals will be able to overlap with those of new atoms so permitting rearrangement to a more stable configuration. Many polymerisation reactions are catalysed by transition metals, *e.g.* the Reppe synthesis of cyclo-octatetraene from acetylene in the presence of nickel. These reactions may be explained by the same mechanism. The cyclobutene-butadiene interconversion in the presence of silver has been explained by the same process [147].

The polymerization of acetylenes has been discussed in terms of the bonding scheme involving dp^2 hybridization of the metal ion [138]. It has been argued that acetylene complexes of divalent metals will be inherently unstable (unless sterically or chemically inhibited) since only one of the two ligand π-orbitals is co-ordinated. The unco-ordinated π-orbital will be synergically destabilised allowing it to act as a Lewis base and so take part in reactions impossible for the free acetylene molecule.

8. Infra-red Spectra of Olefin and Acetylene Complexes

Although spectral evidence may only be used qualitatively nevertheless it has often been taken as evidence in assessing the mode of bonding in transition-metal complexes. Correlations have been made between the change in the carbon-carbon bond stretching frequency on co-ordination of an unsaturated ligand to a series of transition metals with the metal-olefin or metal-acetylene bond strength. However, a recent infra-red study of Zeise's salt [133] has suggested that the C=C stretching mode of an olefin often couples with the in-plane scissoring vibration and a further investigation [132] has indicated that the metal-olefin stretching force constant serves as the best measure of the strength of co-ordination, more especially for a series of different olefins with one particular metal. For example, Table 15 lists the platinum-olefin stretching frequencies of various mono-olefin complexes together with the C=C stretching frequencies of the free and co-ordinated olefins; there is no apparent parallel between the lowering of the C=C stretching frequency and the platinum-olefin stretching frequency.

In a recent review Quinn and Tsai [2] tabulated data for a series of ethylene-metal carbonyl complexes and concluded that for a given metal or group in the

Table 15. *The C=C and Pt-olefin stretching frequencies of several mono-olefin platinum complexes*

Complex	ν(C=C) of free olefin	ν(C=C) of coord. olefin	$\Delta\nu$(C=C)	ν(metal-olefin
$K[Pt(C_2H_4)Cl_3]H_2O$	1623	1526	97	407
$K[Pt(C_2D_4)Cl_3]H_2O$	1515	1428	87	387
$K[Pt(C_2H_4)Br_3]H_2O$	1623	1511	112	395
$K[Pt(C_3H_6)Cl_3$	1649	1505	144	393
$K[Pt(trans\text{-}C_4H_8)Cl_3]$	1681	1522	159	387
$K[Pt(cis\text{-}C_4H_8)Cl_3]$	1672	1505	167	405
$trans\text{-}[Pt(C_2H_4)NH_3 \cdot Cl_2]$	1623	1521	102	383
$trans\text{-}[Pt(C_2H_4)NH_3 \cdot Br_2]$	1623	1517	106	383

Periodic Table, a parallelism existed between the direction of change of ν_{CO} and $\nu_{C=C}$ when a change occurred in the complex. Since a change in the d-electron donor capacity of the metal would be felt by both the olefin and the carbonyl ligands, then the simultaneous decrease in both ν_{CO} and $\nu_{C=C}$ implies a greater d–π^* overlap for both types of ligands. A clear difference was listed between the silver(I)-ethylene complex [50] and ethylene complexes of other transition metals. Comparatively small decreases in $\nu_{C=C}$ have been noticed for several silver(I) complexes; for example, Raman studies of silver-olefin complexes showed a general lowering of $\nu_{C=C}$ of approximately 50–70 cm^{-1}, while monosubstituted acetylenes showed a lowering of the triple-bond frequency by 100 cm^{-1} [150]. Also silver(I), copper(I) and platinum(II) complexes of allyl pyridines show $\Delta\nu_{C=C}$ values of 58, 87 and 187 cm^{-1} [150]. The similar small changes for copper(I)-olefin complexes has suggested that the σ-component of the metal-olefin bond may be more important than the π-component in these Group Ib metal complexes. Whether this also applies to gold(I) complexes is questionable since complex formation results in a lowering of $\nu_{C=C}$ of about 115 cm^{-1} for 1-olefins and 125–135 cm^{-1} for cyclic olefins [151].

Silver complexes of olefins showed an increased shift to lower frequencies of the C=C stretching vibration which was dependent upon increased alkyl substitution at the double bond [127]. For mono-, di-, and tri- alkyl substitution the magnitude is about 55, 63 and 70 cm^{-1} respectively, reflecting the increased overlap of the olefin π-orbital with the silver ion orbital due to the increasing π-basicity of the olefin. Furthermore, Quinn and Glew [50] were able to propose a linear correlation between the decrease in $\nu_{C=C}$ for the complexed olefin and both the ionization potential (which decreases with increasing olefin basicity) and ΔH^0_D, the enthalpy change for complex dissociation to solid silver salt and one mole of gaseous olefin.

In olefin and acetylene complexes of transition metals, in contrast to those of silver(I) and possibly copper(I), the π-component of the co-ordinate bond is considered to predominate over the σ-component. This has been inferred from the large shifts in $\nu_{C=C}$ on co-ordination of the unsaturated ligands. Olefins on co-ordination to platinum and palladium show a decrease in $\nu_{C=C}$ of 140 cm^{-1} and 120 cm^{-1} respectively; acetylenes, on the other hand, show decrease of 250 cm^{-1} and 400 – 575 cm^{-1} on co-ordination to platinum(II) and platinum (0) respectively [152].

While palladium(II) complexes of acetylenes have not been synthesised, since rapid polymerisation occurs, those of palladium(0) have been prepared and the $\nu_{C\equiv C}$ is compared with other complexes in Table 16. There is obviously a wide variation in $\nu_{C\equiv C}$ and hence in bond order when acetylenes are co-ordinated. It has been assumed that the lower the value of $\nu_{C\equiv C}$, the greater was the degree of sp^2 hybridisation and hence the stronger the metal acetylene bond.

Inevitably, a great deal of attention has been paid to the analysis of the spectra of Zeise's salt, $K[PtCl_3C_2H_4]$, and Zeise's dimer $[Pt_2Cl_2C_2H_4]_2$.

Table 16

Compound	$\nu_{C\equiv C}(cm^{-1})$	Ref.
$CF_3 \cdot C\equiv C \cdot CF_3$	2300 (raman)	a
pi-$C_5H_5Mn(CO)_2C_4F_6$	1919	b
$(\phi_3P)_2RhCl \cdot C_4F_6$	1917	c
$(\phi_3P)_2 \cdot Pd \cdot C_4F_6$	1811, 1838	d
$(pi$-$C_5H_5)_2 \cdot V \cdot C_4F_6$	1800	e
$(\phi_3P_3)_2 \cdot Pt \cdot C_4F_6$	1775	b
$(\phi_3P)_2 \cdot Ir(CO)CL \cdot C_4F_6$	1773	f

a Miller, F. A., Bauman, R. P.: J. Chem. Phys. *22*, 1544 (1954).
b Baston, J. L., Grim, S. D., Wilkinson, G.: J. Chem. Soc. *1963*, 3468.
c Mays, M. J., Wilkinson, G.: J. Chem. Soc. *1965*, 6629.
d Greaves, E. O., Maitlis, P. M.: J. Organometal. Chem. (Amsterdam) *6*, 104 (1966).
e Tsumura, R., Hagihara, N.: Bull. Chem. Soc. Japan *38*, 861, 1901 (1965).
f Parshall, G. W., Jones, F. N.: J. Am. Chem. Soc. *87*, 5356 (1965).

Several infra-red studies have made assignments of the various bonds in the spectrum and the carbon-carbon stretching frequency has always been taken as the weak band near 1520 cm^{-1}. Although this mode has been considered to couple strongly with the CH_2 scissoring mode (making any correlation with bond strength only qualitative) more recent investigations [153)] have proposed that the strong polarised Raman band at about 1240 cm^{-1} to be the true assignment for the carbon-carbon stretching frequency. On deuteration this increases by 110 cm^{-1}. The platinum-carbon stretching frequencies (405 cm^{-1} and 493 cm^{-1}) indicate the ethylene to be tightly bound to the platinum, contrasting to previous conclusions.

Chatt and Duncanson [131)], after analysis of the infra-red spectrum, proposed a structure for Zeise's salt where the ethylene was situated at a right-angle to the plane of the platinum atom and the three chlorine atoms. Hiraishi [153)] has suggested that the assignment of the carbon-carbon stretching frequency to 1520 cm^{-1} still indicated a large amount of double bond character to be present in the bond since the $\nu_{C=C}$ for ethylene is 1623 cm^{-1} and ν_{C-C} for ethane is 993 cm^{-1}. This indication of a weak platinum-olefin bond contrasted with evidence from platinum-ethylene stretching vibrations, and consequently he re-assigned the carbon-carbon stretching frequency to about 1240 cm^{-1}. This decrease in the nature of the double bond in Zeise's salt and the observation of this platinum-ethylene vibrations indicate that d–p_π overlap plays a not unimportant part in the co-ordinate bond.

F. Conclusions

All available evidence points to the compound nature of the metal-olefin *pi*-bond. The MO descriptions outlined in Sect. E. all suggest both σ- and π-contributions to the total bonding and the presence of both forward and back bonding finds support in stability constants, complex geometry and bond lengths and in theoretical calculations. As a general rule the σ-component appears to predominate although the π-component is important in complexes of zerovalent metals or with ligands containing strongly electron-attracting groups.

The synergic importance of limited π-overlap on the metal-olefin σ-bond should not be underestimated. The illustration F.I. demonstrates that a small

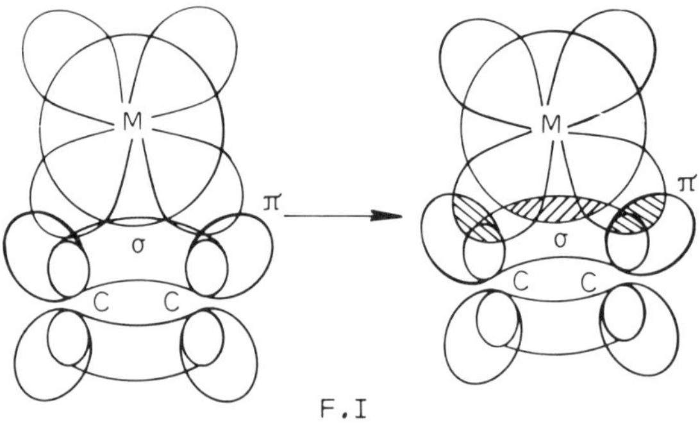

F.I

increase in π-overlap, causing a small bond shortening, will have a vastly enhanced effect on the σ-overlap. Hence, while the σ-overlap may largely account for the thermodynamic bond energy the contribution from the π-overlap is extremely important.

In practice a continuous range of bond-types from almost pure σ to almost pure π-bonding is to be expected. Factors determining the position within this range have been discussed by Greaves, Lock and Maitlis and a simple picture has evolved [3]. Three criteria can be given if significant bond formation is to occur.

i) Orbitals of the same symmetry must exist on both the metal and the ligand.
ii) The orbitals are able to overlap significantly.
iii) The energy of the overlapping orbitals must be similar.

Assuming criteria i) and ii) are satisfied, iii) will be the deciding factor. The various ligand orbitals are shown in F. II. The metal valence electrons can fall into one of a number of regions. In region A the orbitals lie at a comparable

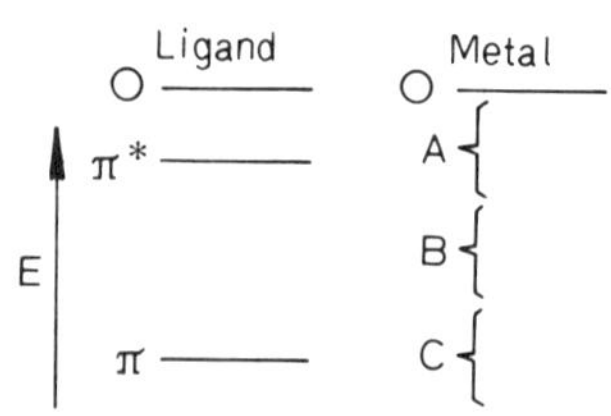

F.II

energy to the ligand π^* orbitals, in region B they are between the π and π^* energies and in region C they are more comparable with the π orbital. In region A complex formation will involve substantial M→ligand charge transfer (π-bonding); in region B there is likely to be substantial charge transfer in both the M→L and the M←L bonds leading to a high M–L bond order and encouraging the ligand to take on the geometry of an excited state. In region C bonding will be more nearly M←L σ-bonding with little contribution from back co-ordination. As a ligand TCNE will contain π^* orbitals of reasonably low energy while Pt(0) will have readily ionized valence electrons. Hence the Pt(O)–TCNE *pi*-bond would be likely to fall into region A giving a bond of largely π-character. On the other hand the Pt(II)–C_2Me_4 π-bond would be nearer to region C, giving a bond of largely σ-character. Since regions A–C represent a continuous graduation of bond type metal-olefin and metal-acetylene bonds are best represented as in F.III.

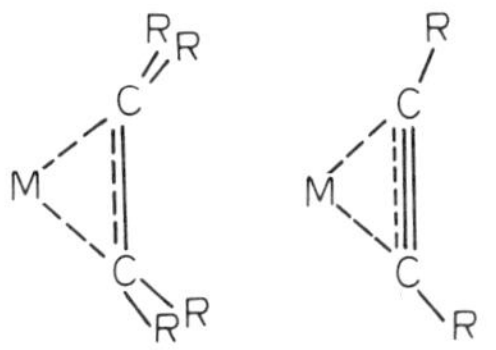

F.III

G. References

1) Hartley, F. R.: Chem. Rev. *69,* 799 (1969).
2) Quinn, H. W., Tsai, J. H.: Advan. Inorg. Chem. Radiochem. *12,* 217 (1970).
3) Greaves, E. O., Lock, C. J. L., Maitlis, P. M.: Can. J. Chem. *46,* 3879 (1968).
4) Panattoni, C., Bombieri, G., Belluco, U., Baddley, W. H.: J. Am. Chem. Soc. *90,* 798 (1968).
5) Raper, G., McDonald, W. S.: Chem. Commun. *1970,* 655.
6) Racanelli, P., Pantini, G., Immirzi, A., Allegra, G., Porri, L.: Chem. Commun. *1969,* 361.
7) Davies, G. R., Hewertson, W., Mais, R. H. B., Owston, P. G.: Chem. Commun. *1967,* 423.
8) Mills, O. S., Shaw, B. W.: Acta Cryst. *18,* 562 (1965).
9) Dahl, L. F., Smith, D. L.: J. Am. Chem. Soc. *84,* 2451 (1962).
10) Ahrland, S., Chatt, J., Davies, N. R.: Quart. Rev. (London) *12,* 265 (1958).
11) Eberz, W. F., Wedge, H. J., Yost, D. M., Lucas, H. J.: J. Am. Chem. Soc. *59,* 45 (1937).
12a) Winstein, S., Lucas, H. J.: J. Am. Chem. Soc. *60,* 836 (1938).
12b) Lucas, H. J., Moore, R. S., Pressman, D.: J. Am. Chem. Soc. *65,* 227 (1943).
12c) Lucas, H. J., Billmeyer, F. W., Pressman, D.: J. Am. Chem. Soc. *65,* 230 (1943).
12d) Hepner, F. R., Trueblood, K. N., Lucas, H. J.: J. Am. Chem. Soc. *74,* 1333 (1952).
12e) Trueblood, K. N., Lucas, H. J.: J. Am. Chem. Soc. *74,* 1338 (1952).
13) Eley, D. D.: Cationic Polymerisation, p. 6. Cambridge: Heffer and Sons 1953.

14) Dorsey, W. S., Lucas, H. J.: J. Am. Chem. Soc. *78,* 1665 (1956).
15a) Kofahl, R. E., Lucas, H. J.: J. Am. Chem. Soc. *76,* 3931 (1954).
15b) Andrews, L. J., Keefer, R. M.: J. Am. Chem. Soc. *71,* 3644 (1949).
16) Helmkamp, G. K., Carter, F. L., Lucas, H. J.: J. Am. Chem. Soc. *79,* 1306 (1957).
17) Dewar, M. J. S.: Bull. Soc. Chim. France *18,* C71 (1951).
18) Gardner, P. D., Brandon, R. L., Nix, N. J.: Chem. Ind. *1958,* 1363.
19) Traynham, J. G., Olechowski, J. R.: J. Am. Chem. Soc. *81,* 571 (1959).
20) Traynham, J. G., Sehnert, M. L.: J. Am. Chem. Soc. *78,* 4024 (1956).
21) Muhs, M. A., Weiss, F. T.: J. Am. Chem. Soc. *84,* 4697 (1962).
22) Fueno, T., Okuyama, T., Deguchi, T., Furukawa, J.: J. Am. Chem. Soc. *87,* 170 (1965).
23) Hosoya, H., Nagakura, S.: Bull. Chem. Soc. Japan *37,* 249 (1964).
24) Fukui, K., Imamura, A., Yonezawa, T., Nagata, C.: Bull. Chem. Soc. Japan *34,* 1076 (1961).
25) Fueno, T., Okuyama, T., Furukawa, J.: Bull. Chem. Soc. Japan *39,* 2094 (1966).
26) Fueno, T., Kajimoto, O., Furukawa, J.: Bull. Chem. Soc. Japan *41,* 782 (1968).
27) Fueno, T., Kajimoto, O., Okuyama, T.: Bull. Chem. Soc. Japan *41,* 785 (1968).
28) Okuyama, T., Fueno, T.: Bull. Chem. Soc. Japan *42,* 3106 (1969).
29) Cvetanovic, R. J., Duncan, F. J., Falconer, W. E., Irwin, R. S.: J. Am. Chem. Soc. *87,* 1827 (1965).
30) Temkin, O. N., Flid, R. M., Malakhov, A. I.: Kinetika i Kataliz *3,* 915 (1962).
31) Temkin, O. N., Ginzburg, A. G., Flid, R. M.: Kinetika i Kataliz *5,* 221 (1964).
32) Brandt, P.: Acta Chem. Scand. *13,* 1639 (1959).
33) Petrov, A. N., Temkin, O. N., Bogdanov, M. I.: Neftekhimiya *9,* 729 (1969).
34) Genkin, A. N., Boguslavskaya, B. I.: Neftekhimiya *5,* 897 (1965).
35) Rowland, R. L., Kluchesky, E. F.: J. Am. Chem. Soc. *73,* 5490 (1951).
36) Gil-Av, E., Herling, J.: J. Phys. Chem. *66,* 1208 (1962).
37) Hosoya, H., Nagakura, S.: Bull. Chem. Soc. Japan *37,* 249 (1964).
38) Smith, B., Ohlson, R.: Acta Chem. Scand. *16,* 351 (1962).
39) Shopov, D., Andrev, A.: Compt. Rend. Acad. Bulgare Sci. *19,* 499 (1966).
40a) Featherstone, W., Sorrie, A. J. S.: J. Chem. Soc. *1964,* 5235.
40b) Baker, B. B.: Inorg. Chem. *3,* 200 (1964).
41) Nichols, P. L.: J. Am. Chem. Soc. *74,* 1091 (1952).
42) Resnik, R. K., Douglas, B. E.: Inorg. Chem. *2,* 1246 (1963).
43) Franzus, B., Baird, W. C., Snyder, E. I., Surridge, J. H.: J. Org. Chem. *32,* 2845 (1967).
44) Gray, D., Wies, R. A., Closson, W. D.: Tetrahedron Letters *1968,* 5639.
45) Hartley, F. R., Venanzi, L. M.: J. Chem. Soc. (A) *1967,* 333.
46) Pettit, L. D., Sherrington, C.: J. Chem. Soc. (A) *1968,* 3078.
47) Barnes, D. S., Ford, G. J., Pettit, L. D., Sherrington, C.: Chem. Comm. *1971,* 690.
48) Menon, B. C., Pincock, R. E.: Can. J. Chem. *47,* 3327 (1969).
49) Clever, H. L., Baker, E. R., Hale, W. R.: J. Chem. Eng. Data *15,* 411 (1970).
50) Quinn, H. W., Glew, D. N.: Can. J. Chem. *40,* 1103 (1962).
51) Tamara, K., Sano, M., Tatsuoka, K.: Bull. Chem. Soc. Japan *36,* 1366 (1963).
52) Francis, A. W.: J. Am. Chem. Soc. *73,* 3709 (1951).
53) Kraus, J. W., Stern, E. W.: J. Am. Chem. Soc. *84,* 2893 (1962).
54a) Tropsch, H., Mattox, W. J.: J. Am. Chem. Soc. *57,* 1102 (1935).
54b) Gilliland, E. R., Seebold, J. E., Fitzhugh, J. R., Morgan, P. S.: J. Am. Chem. Soc. *61,* 1960 (1939).
55) Gilliland, E. R., Bliss, H. L., Kip, C. E.: J. Am. Chem. Soc. *63,* 2088 (1941).
56) Slade, P. E., Jonassen, H. B.: J. Am. Chem. Soc. *79,* 1277 (1957).
57a) Keefer, R. M., Andrews, L. J.: J. Am. Chem. Soc. *71,* 1723 (1949).

[57b)] Keefer, R. M., Andrews, L. J., Kepner, R. E.: J. Am. Chem. Soc. *71*, 3906 (1949).
[58a)] Andrews, L. J., Keefer, R. M.: J. Am. Chem. Soc. *70*, 3261 (1948).
[58b)] Andrews, L. J., Keefer, R. M.: J. Am. Chem. Soc. *71*, 2379 (1949).
[58c)] Keefer, R. M., Andrews, L. J., Kepner, R. E.: J. Am. Chem. Soc. *71*, 2381 (1949).
[59)] Basolo, F., Perason, R. G.: Mechanisms of Inorganic Reactions, Chap. 8. New York: J. Wiley and Sons.
[60)] Temkin, O. N., Flid, R. M., German, E. D., Onishehenko, T. A.: Kinetika i Kataliz *2*, 205 (1961).
[61)] Osterlöf, J.: Acta Chem. Scand. *4*, 374 (1950).
[62)] Van der Hende, J. H., Baird, W. C.: J. Am. Chem. Soc. *85*, 1009 (1963).
[63)] Trebellas, J. C., Oleglowski, J. R., Jonassen, H. B.: Inorg. Chem. *4*, 1818 (1965).
[64)] Haight, H. L., Doyle, J. R., Baenziger, N. C., Richards, G. F.: Inorg. Chem. *2*, 1301 (1963).
[65)] Manahan, S. E.: Inorg. Chem. *5*, 2063 (1966).
[66)] Manahan, S. E.: Inorg. Chem. *5*, 482 (1966).
[67)] Harvilchuck, J. M., Aikens, D. A., Murray, R. C.: Inorg. Chem. *8*, 539 (1969).
[68)] Baenziger, N. C., Richards, G. F., Doyle, J. R.: Acta Cryst. *18*, 924 (1965).
[69a)] Schrauzer, G. N.: Cem. Ber. *94*, 643 (1961).
[69b)] Schrauzer, G. N., Eichler, S.: Chem.Ber. *95*, 260 (1962).
[70)] Kawaguchi, S., Ogura, T.: Inorg. Chem. *5*, 844 (1966).
[71)] Treger, Yu. A., Flid, R. M., Antanova, L. V., Spektor, S. S.: Russ. J. Phys. Chem. *39*, 1515 (1965).
[72)] Joy, J. R., Orchin, M.: J. Am. Chem. Soc. *81*, 305 (1959).
[73)] Joy, J. R., Orchin, M.: J. Am. Chem. Soc. *81*, 310 (1959).
[74)] Shupack, S. I., Orchin, M.: J. Am. Chem. Soc. *86*, 586 (1964).
[75)] Denning, R. G., Hartley, F. R., Venanzi, L. M.: J. Chem. Soc.(A) *1967, 324.*
[76)] Denning, R. G., Venanzi, L. M.: J. Chem. Soc. (A) *1967*, 336.
[77)] Denning, R. G., Hartley, F. R., Venanzi, L. M.: J. Chem. Soc. (A) *1967*, 328.
[78)] Hartley, F. R., Venanzi, L. M.: J. Chem. Soc. (A) *1967*, 330.
[79)] Milburn, R. M., Venanzi, L. M.: Inorg. Chim. Acta *2*, 97 (1968).
[80)] Spagna, R., Venanzi, L. M., Zambonelli, L.: Inorg. Chim. Acta *4*, 475 (1970).
[81)] Spagna, R., Venanzi, L. M., Zambonelli, L.: Inorg. Chim. Acta *4*, 283 (1970.
[82)] Ahrland, S.: Struct. Bonding *5*, 118 (1968).
[83a)] Elding, L. I., Leden, I.: Acta Chem. Scand. *20*, 706 (1966).
[83b)] Drougge, L., Elding, L. I., Gustafson, L.: Acta Chem. Scand. *21*, 1647 (1967).
[84)] Jorgensen, C. K.: Inorg. Chem. *3*, 1201 (1964).
[85)] McMane, D. G.: Dissertation Abstr. B*28*, 3629 (1968).
[86)] Chatt, J., Shaw, B. L., Williams, A. A.: J. Chem. Soc. *1962*, 3269.
[87a)] Allen, A. D., Cook, C. D.: Can. J. Chem. *41*, 1235 (1963).
[87b)] Allen, A. D., Cook, C. D.: Can. J. Chem. *42*, 1063 (1964).
[88)] Cook, C. D., Danyluk, S. S.: Tetrahedron *19*, 177 (1963).
[89)] Chatt, J., Rowe, G. A., Williams, A. A.: Proc. Chem.Soc. *1957*, 208.
[90)] Baddely, W. H., Venanzi, L. M.: Inorg. Chem. *5*, 33 (1966).
[91)] See Refs. 139, 140.
[92a)] Moiseev, I. I., Vargaftik, M. N., Syrkin, Ya. K.: Dokl. Akad. Nauk SSSR *152*, 147 (1963); C. A. *60*, 184d (1964).
[92b)] Pestrikov, S. V., Moiseev, I. I.: Izv. Akad. Nauk SSSR, Ser. Khim. *1965*, 349. C. A. *62*, 16018d (1965).
[92c)] Pestrikov, S. V., Moiseev, I. I., Romanova, T. N.: Zh. Neorgan. Khim. *10*, 2203 (1965); C. A. *63*, 15836 (1965).

92d) Pestrikov, S. V., Moiseev, I. I., Tsivilikhovskaya, B. A.: Zh. Neorgan. Khim. *11,* 1742 (1966); C. A. *65,* 13757c (1966).

92e) Pestrikov, S. V., Moiseev, I. I., Sverzh, L. M.: Zh. Neorgan. Khim. *11,* 2081 (1966); C. A. *66,* 45991 (1967).

93a) Henry, P. M.: J. Am. Chem. Soc. *86,* 3246 (1964).

93b) Henry, P. M.: J. Am. Chem. Soc. *88,* 1595 (1966).

94) Cramer, R.: J. Am. Chem. Soc. *89,* 4621 (1967).

95) Osborn, J. A., Jardine, F. H., Young, J. F., Wilkinson, G.: J. Chem. Soc. (A) *1966.* 1771.

96) Gustorf, E. K. von, Henry, M. C., McAdoo, D. J.: Liebigs Ann. Chem. *707,* 190 (1967).

97) Cetini, G., Gambino, O., Stanghellini, P. L., Vaglio, G. A.: Inorg. Chem. *6,* 1225 (1967).

98) Allen, E. R., Cartlidge, J., Taylor, N. M., Tipper, C. F. H.: J. Phys. Chem. *63,* 1437 (1959).

99) Kreevoy, M. M., Schalager, L. L., Ware, J. C.: Trans Faraday Soc. *58,* 2433 (1962).

100) Kawai, I., Nakajiama, R., Hara, T.: Bull. Chem. Soc. Japan *43,* 749 (1970).

101) Temkin, O. N., Flid, R. M., Malakhov, A. I.: Kinetika i Kataliz *4,* 270 (1963).

102) Hamilton, W. C.: Acta Cryst. *25A,* 5172 (XIV–46) (1969).

103) Black, M., Mais, R. H. B., Owston, P. G.: Acta Cryst. *25B,* 1753 (1969).

104) Klanderman, K. A.: Diss. Abs. *25,* 6253 (1965).

105) Baenziger, N. C., Medrud, R. C., Doyle, J. R.: Acta Cryst. *18,* 237 (1965).

106) Ibers, J. A., Snyder, R. G.: Acta Cryst. *15,* 923 (1962).

107) Baenziger, N. C., Haight, H. C., Alexander, R., Doyle, J. R.: Inorg. Chem. *5,* 1399 (1966).

108) Cotton, F. A.: Chem. Brit. *4,* 345 (1968).

109) Bennett, M. A., Saxby, J. D.: Inorg. Chem. *7,* 321 (1968).

110) Ziegler, M.: Z. Anorg. Allg. Chim. *355,* 12 (1967).

111a) Dierks, H., Dietrich, H.: Z. Krist. *122,* 1 (1965).

111b) Hende, J. H., Baird, W. C.: J. Am. Chem. Soc. *85,* 1009 (1963).

112) Malone, J. F., McDonald, W. S., Shaw, B. L., Shaw, G.: Chem. Commun. *1968,* 869.

113) Hitchcock, P. B., McPartlin, Mary, Mason, R.: Chem. Commun. *1969,* 1367.

114) McGinnety, J. A., Ibers, J. A.: Chem. Commun. *1968,* 235.

115a) Cook, C. D., Koo, C. H., Nyburg, S. C., Shiomi, M. T.: Chem. Commun. *1967,* 426.

115b) Dreissig, W., Dietrich, H.: Acta Cryst. *24B,* 108 (1968).

116) McGinnety, J. A., Mays, M. J.: Ann. Rep. Chem. Soc. *65,* 381 (1968).

117) Kashiwagi, T., Yasuoky, N., Kasai, N., Kukudo, M.: Chem. Commun. *1969,* 317.

118) Mason, R., Rae, A. I. M.: J. Chem. Soc(A) *1970,* 1767.

119) Johnson, M. D., Mayle, C.: Chem. Commun. *1969,* 192.

120a) Raper, G., McDonald, W. S.: Chem. Commun. *1970,* 655.

120b) McDonald, W. S., Mann, B. E., Raper, G., Shaw, B. L., Shaw, G.: Chem. Commun. *1969,* 1254.

121) Malone, J. F., McDonald, W. S.: J. Chem. Soc. (A) *1970,* 3124.

122) Mason, R., Robertson, G. B., Whimp, P. O.: J. Chem. Soc. (A) *1970,* 535.

123) Nyholm, R. S.: Suomen Kemistilehti *42B,* 165 (1969).

124) Glanville, J. O., Stewart, J. M., Grim, S. O.: Organometal. Chem. Rev. *7,* 9 (1967).

125) Bailey, N. A., Mason, R.: J. Chem. Soc.(A) *1968,* 1293.

126) Gowling, E. W., Kettle, S. F. A., Sharples, G. M.: Chem. Commun. *1968,* 21.

127) Quinn, H. W., McIntyre, J. S., Peterson, D. J.: Can. J. Chem. *43,* 2896 (1965).

128) Schug, J. C., Martin, R. J.: J. Phys. Chem. *66,* 1554 (1962).

129) Parker, R. G., Roberts, J. D.: J. Am. Chem. Soc. *92,* 743 (1970).

130) Gee, D., Wan, J. K. S.: Chem. Commun. *1970,* 641.

131) Chatt, J., Duncanson, L. A.: J. Chem. Soc. *1953,* 2939.
132) Grogan, M. J., Nakamoto, K.: J. Am. Chem. Soc. *90,* 918 (1968).
133) Grogan, M. J., Nakamoto, K.: J. Am. Chem. Soc. *88,* 5454 (1966).
134) Moore, J. W.: Acta Chem. Scand. *20,* 1154 (1966).
135) Bodner, G. M., Storhoff, B. N., Doddrell, D., Todd, L. J.: Chem. Commun. *1970,* 1530.
136) Cook, C. D., Wan, K. Y.: J. Am. Chem. Soc. *92,* 2595 (1970).
137) Nelson, J. M., Wheelock, K. S., Cusachs, L. C., Jonassen, H. B.: J. Am. Chem. Soc. *91,* 7005 (1969).
138) Wheelock, K. S., Nelson, J. M., Cusachs, L. C., Jonassen, H. B.: J. Am. Chem. Soc. *92,* 5110 (1970).
139) Cramer, R., Kline, J. B., Roberts, J.D.: J. Am. Chem. Soc. *91,* 2519 (1969); *86,* 217 (1964); *89,* 5377 (1967).
140) Holloway, C. E., Hulley, G., Johnson, B. F. G., Lewis, J.: J. Chem. Soc. (A) *1969,* 53.
141) Maricic, S., Redpatch, C. R., Smith, J. A. S.: J. Chem. Soc. *1963,* 4905.
142) Mason, R.: Nature *217,* 543 (1968).
143) McWeeny, R., Mason, R., Towl, A. D. C.: Diss. Faraday Soc. *47,* 20 (1969).
144) Kroto, H. W., Santry, D. P.: J. Chem. Phys. *47,* 792 (1967).
145) Mason, R., Robertson, G. B.: J. Chem. Soc. (A) *1969,* 492.
146) Mango, F. D., Schachtschneider, J. M.: J. Am. Chem. Soc. *89,* 2484 (1967).
147) Pettit, R., Sugahara, M., Wristers, J., Merk, W.: Diss. Faraday Soc. *47,* 71 (1969).
148) Baddley, W. H.: J. Am. Chem. Soc. *90,* 3705 (1968).
149) Volger, H. C., Gaasbeek, M. M. P., Hogeveen, H., Vrieze, K.: Inorg. Chim. Acta *3,* 145 (1969).
150) Taufen, H. J., Murray, M. J., Cleveland, F. F.: J. Am. Chem. Soc. *63,* 3500 (1941); Yingst, R. E., Douglas, B. E.: Inorg. Chem. *3,* 1177 (1964).
151) Huettel, R., Reinheimer, H.: Chem. Ber. *99,* 462, 2778 (1966).
152) Greaves, E. O., Maitlis, P. M.: J. Organometal. Chem. *6,* 104 (1966).
153) Hiraishi, J.: Spectrochim. Acta *25A,* 749 (1969).

Received March 12, 1971

Ringliganden-Verdrängungsreaktionen von Aromaten-Metall-Komplexen

Prof. Dr. Helmut Werner
Anorganisch-chemisches Institut der Universität Zürich (Schweiz)

Inhalt

1. Einführung

Die Reaktionen von Aromaten-Metall-Komplexen, d.h. von Komplexen, in denen ein oder mehrere aromatische Ringliganden C_nH_n zentrosymmetrisch an ein Übergangsmetall gebunden sind, können in zwei große Gruppen eingeteilt werden:

1) in Umsetzungen, bei denen die Ring-Metall-Bindung (oder die Ring-Metall-Bindungen) erhalten bleibt, und
2) in solche, bei denen der Ligand C_nH_n (oder die Liganden C_nH_n) durch einen oder mehrere andere Koordinationspartner verdrängt wird.

Zu der ersten Gruppe dieser Umsetzungen gehören z.B. die Substitutionsreaktionen des *Ferrocens,* der Stammsubstanz der Aromaten-Metall-Komplexe. Ihrer Bereitwilligkeit, mit zahlreichen elektrophilen und nucleophilen Agenzien – ähnlich wie das Benzol („*benzene*") – zu reagieren, verdankt diese Verbindung bekanntlich ihren Namen.

Die vorliegende Übersicht ist der zweiten Gruppe der genannten Umsetzungen gewidmet. Sie gelten im allgemeinen als weniger charakteristisch für C_nH_n-M-Komplexe, vor allem wohl, weil in vielen Fällen eine relativ hohe Aktivierungsenergie erforderlich ist. Dennoch bilden die Ringliganden-Verdrängungsreaktionen einen wichtigen Aspekt der Chemie der Aromaten-Metall-Komplexe. Die Ausführungen sollen zeigen, daß bereits viele interessante Ergebnisse vorliegen, die nicht nur aus dem Blickwinkel des präparativ arbeitenden Chemikers sondern auch aus bindungstheoretischer und reaktionsmechanistischer Sicht Beachtung verdienen.

Zum stofflichen Umfang sei folgendes gesagt: Es werden in erster Linie Reaktionen von C_5H_5–M- und C_6H_6–M-Komplexen besprochen. Über analoge Umsetzungen von C_7H_7-M- und C_8H_8-M-Verbindungen (C_8H_8 als Anion $C_8H_8^{2-}$ gebunden) ist bis jetzt nichts bekannt. Obwohl Cyclobutadien definitionsgemäß *kein* Aromat, d.h. kein monocyclisches Ringsystem mit $(4n + 2)$ π-Elektronen ist, werden Liganden-Übertragungsreaktionen von C_4H_4-M-Komplexen unter 4. erwähnt. Sie sind in vielem den entsprechenden Umsetzungen von C_5H_5-M- und C_6H_6-M-Verbindungen an die Seite zu stellen. Nicht erörtert werden Ligandenverdrängungsreaktionen von Metallkomplexen cyclischer Oligoolefine, wie z.B. Cycloheptatrien oder Cyclooctatetraen; hierüber liegen bereits mehrere zusammenfassende Arbeiten vor [1]. Auf vergleichende mechanistische Aspekte der Umsetzungen von C_6H_6-M- und C_7H_8-M-Verbindungen wird unter 3. kurz hingewiesen. Ebenfalls nicht besprochen werden Reaktionen, bei denen die Spaltung einer Ring-Metall-Bindung durch Einwirkung eines Oxydationsmittels wie z.B. Ag^+ oder Ce^{4+} erfolgt. Hierbei ist nicht der Ligandenaustausch sondern der Redox-Vorgang reaktionsbestimmend.

Die folgenden *Abkürzungen* werden verwendet:

Me	$= CH_3$,
Et	$= C_2H_5$,
But	$= n - C_4H_9$,
Ph	$= C_6H_5$,
Ar	= Aromat,
dipy	= 2.2'-Dipyridyl,
tripy	= 2.2'.2"-Tripyridyl,
phen	= 1.10-Phenanthrolin.

2. Präparative Ergebnisse

2.1. Reaktionen von Dicyclopentadienylnickel

Die recht zahlreich untersuchten Umsetzungen von Dicyclopentadienylnickel mit Lewis-Basen (L) verdienen insofern an den Anfang gestellt zu werden, als in ihrem Verlauf – in Abhängigkeit von L und den verwendeten Reaktionsbedingungen – unterschiedliche Verbindungstypen entstehen können. Es sind drei Reaktionsmöglichkeiten zu unterscheiden:

a) Spaltung *beider* $\pi-C_5H_5$–Ni-Bindungen und Bildung tetraedrischer Nickel(0)-Komplexe NiL_4 (L = einzähniger Ligand) bzw. NiL'_2 (L' = zweizähniger Ligand);

b) Spaltung nur *einer* $\pi-C_5H_5$–Ni-Bindung und Bildung von ein- oder zweikernigen Monocyclopentadienyl-Nickel-Komplexen;

c) Addition von L (z.B. C_2F_4) an einen der Cyclopentadienylringe und Umwandlung desselben in ein π-Allyl-System.

Die Ligandenverdrängung nach a) wurde erstmals bei der Umsetzung von $Ni(C_5H_5)_2$ mit CO beobachtet. Bei Versuchen, das nach der Edelgasregel als stabil zu erwartende $[C_5H_5NiCO]_2$ gemäß (1) zu erhalten, entstand auch unter relativ schonenden Bedingungen nicht der gesuchte Zweikernkomplex sondern $Ni(CO)_4$ [2].

$$2\,Ni(C_5H_5)_2 + 2\,CO \longrightarrow [C_5H_5NiCO]_2 + C_{10}H_{10} \qquad (1)$$

Wie nachfolgende Untersuchungen zeigten, ist die Bildung von NiL_4 oder NiL'_2 nach (2) oder (3) auch bei der Einwirkung zahlreicher anderer n-Donoren der vorherrschende Reaktionsverlauf.

$$Ni(C_5H_5)_2 + 4\,L \longrightarrow NiL_4 + C_{10}H_{10} \qquad (2)$$

$$Ni(C_5H_5)_2 + 2\,L' \longrightarrow NiL'_2 + C_{10}H_{10} \qquad (3)$$

Bisher wurden auf diesem Wege Komplexe NiL_4 mit

$$L = CNR, PR_3, P(OR)_3, PF_3, PF_2R, PFR_2, P_4S_3$$

und Komplexe NiL'_2 mit

L' = dipy, tripy, phen, 1.2-Bis(diphenylphosphino)äthan und
1.1.1-Tris(diphenylphosphinomethyl)äthan

synthetisiert. Die verwendeten Liganden, die Reaktionsbedingungen und Ausbeuten sind in Tabelle 1 zusammengestellt.

Aus der Übersicht geht hervor, daß $Ni(C_5H_5)_2$ eine sehr geeignete Ausgangssubstanz für die Synthese von Nickel(0)-Komplexen ist. Das in organischen Solvenzien im allgemeinen noch besser lösliche Di(methylcyclopentadienyl)nickel hat ebenfalls Verwendung gefunden [5]. Gegenüber $Ni(CO)_4$ – das sonst häufig zur Darstellung von NiL_4 oder NiL'_2 benutzt wird – besitzt $Ni(C_5H_5)_2$ und auch $Ni(C_5H_4Me)_2$ den Vorteil, daß die Ligandenverdrängung nach (2) oder (3) zu einem einzigen Verbindungstyp und nicht zu einem Produktgemisch führt; zudem ist $Ni(C_5H_5)_2$ leichter zu handhaben und weniger giftig als $Ni(CO)_4$.

Bei der Bildung von NiL_4 oder NiL'_2 nach (2) bzw. (3) tritt formal eine Reduktion von Ni(II) zu Ni(0) ein. Dabei ist offensichtlich von Bedeutung, daß die in Tabelle 1 angegebenen Liganden L und L' als π-Akzeptoren fungieren können, d.h. über energetisch günstig liegende, freie Orbitale verfügen. Trifft dies nicht zu, wie z.B. für L = NH_3, so erfolgt Substitution unter Beibehalt der Oxydationszahl +II: $Ni(C_5H_5)_2$ reagiert mit flüssigem NH_3 zu salzartigem $[Ni(NH_3)_6](C_5H_5)_2$ [4]. Die Umsetzung mit KCN in NH_3 nach (4) verläuft analog.

$$Ni(C_5H_5)_2 + 4\,KCN \longrightarrow K_2[Ni(CN)_4] + 2\,KC_5H_5 \qquad (4)$$

Ähnlich wie $Ni(C_5H_5)_2$ reagieren $C_5H_5NiC_3H_5$ und das homologe $C_5H_5PdC_3H_5$ mit Phosphinen, Phosphiten und Isonitrilen [11,12]. Es erfolgt auch hier – für den *Palladium*-Komplex bereits bei 0 °C – eine Verdrängung der beiden π-gebundenen Liganden. Mit Halogenverbindungen wie z.B. $AsCl_3$ [11], $AlCl_3$, $FeCl_2$, $HgCl_2$ oder HCl [13] reagiert $C_5H_5PdC_3H_5$ ausschließlich unter Spaltung der Cyclopentadienyl-Metall-Bindung; es entsteht in allen Fällen der Zweikernkomplex $[C_3H_5PdCl]_2$.

Der Reaktionstyp b), d.h. die Spaltung nur *einer* π–C_5H_5-Bindung, scheint dann bevorzugt zu sein, wenn die im Primärschritt aus einem Molekül $Ni(C_5H_5)_2$ und einem Molekül L gebildete Zwischenverbindung kinetisch genügend stabil ist und nicht sofort mit überschüssigem L weiterreagiert. Die Ergebnisse der Untersuchungen von Ustynyuk und Mitarbeitern [14,15] sowie von van den Akker und Jellinek [16] können hierfür als Beleg dienen.

Tabelle 1. *Reaktionen von Dicyclopentadienylnickel mit Lewis-Basen*

Ligand	Solvens	Temperatur [°C]	Reaktions-dauer	Ausbeute [in %]	Ref.
CO	Dimethylformamid	120			2)
CO (60 at)	–	50–60 a)		100	2)
CNPh	Diäthyläther	0		100	3,4)
dipy	Cyclohexan	60–120 b)		100	4)
phen	Cyclohexan	60–120 b)		100	4)
tripy	Cyclohexan	60–120 b)		100	4)
PPh_3	Cyclohexan	60–120 b)		100	4)
$C_2H_4(PPh_2)_2$	Cyclohexan	60		100	4)
$CH_3C(CH_2PPh_2)_3$	Cyclohexan	60		100	4)
$P(OPh)_3$	(c)	80		96	5)
$P(OCH_2CH_2Cl)_3$	(c)	80		98	5)
$P(O\text{-}p\text{-}C_6H_4OMe)_3$	(c)	80		33	5)
$P(O\text{-}p\text{-}C_6H_4Me)_3$	(c)	80		23	5)
$P(O\text{-}2\text{-}C_6H_{12}OEt)_3$	(c)	80		35	5)
$P(O\text{-}2\text{-}C_6H_{12}Et)$	(c)	80		48	5)
$P(OEt)_3$	Dioxan	40	15^h	82	6)
PF_3	–	60–65 b)	5^d	74	7,8)
PF_2CF_3	–	60 b)	40^h	92	8)
$PF(CF_3)_2$	–	25 b)	4^d	83	8)
PF_2CCl_3	–	25 b)	14^d	82	8)
PF_2NEt_2	Hexan	70	21^h	46	8)
$PF_2NC_5H_{10}$	Hexan	70	21^h	48	8)
PF_2CH_2Cl	–	25	12^h	80	9)
P_4S_3					10)

a) Umsetzung im Autoklav.
b) Umsetzung im Bombenrohr.
c) Umsetzung ohne Solvens oder in Lösungsmitteln wie z.B. Benzol, Cyclohexan, Aethylcyclohexan, Cyclohexen.

Nach den Angaben in Tabelle 1 entsteht bei der Reaktion von $Ni(C_5H_5)_2$ und Triphenylphosphin in Cyclohexan bei 60–120 °C in praktisch quantitativer Ausbeute $Ni(PPh_3)_4$ [4)]. Führt man die gleiche Umsetzung jedoch in CCl_4 bei Raumtemperatur durch, so bildet sich $\pi\text{–}C_5H_5Ni(PPh_3)Cl$ [14)]. Ein Vorschlag zum Verlauf der Reaktion geht davon aus, daß aus $Ni(C_5H_5)_2$ und PPh_3 zunächst ein wenig stabiler π-Cyclopentadienyl-σ-cyclopentadienyl-Komplex $(\pi\text{–}C_5H_5)Ni(\sigma\text{-}C_5H_5)PPh_3$ entsteht, der dann mit CCl_4 (oder bei Verwendung von Benzol als Solvens mit HCl) unter Spaltung der Ni-σ-C_5H_5-Bindung reagiert [14]. (π-C_5H_5)Ni(PPh_3)Cl wird auch erhalten, wenn man Ni$(C_5H_5)_2$ mit Triphenylphosphoniumhalogeniden [PPh_3H]X (X = Br, J) oder einem Gemisch von PPh_3 und HX (X = Cl, Br, J) in Diäthyläther umsetzt. Die entsprechenden Triäthylphosphin-Komplexe sind nach (6) ebenfalls zugänglich [16].

$$Ni(C_5H_5)_2 + [PPh_3H]X \longrightarrow (\pi\text{-}C_5H_5)Ni(PPh_3)X + C_5H_6 \quad (5)$$

$$Ni(C_5H_5)_2 + PR_3 + HX \longrightarrow (\pi\text{-}C_5H_5)Ni(PR_3)X + C_5H_6 \quad (6)$$

Die Tatsache, daß eine zu (5) analoge Reaktion mit [PPh_3Me]J an Stelle von [PPh_3H]J nicht erfolgt, hat zu der Vermutung geführt, daß primär durch Angriff eines Protons auf Ni$(C_5H_5)_2$ das Kation $[C_5H_5NiC_5H_6]^+$ entsteht, das dann mit PR_3 unter Verdrängung des Cyclopentadien-Liganden reagiert [16]. Nach der kürzlich in unserem Arbeitskreis auf zwei verschiedenen Wegen gelungenen Synthese der Verbindung $[C_5H_5NiC_5H_6]BF_4$, deren Umsetzung mit PR_3 (R = But, Ph)

$$[C_5H_5Ni(PR_3)_2]BF_4$$

ergibt [17], scheint dieser Vorschlag richtig zu sein. In diesem Zusammenhang ist bemerkenswert, daß sich auch bei der gemeinsamen Einwirkung von $PBut_3$ und Allylhalogeniden C_3H_5X (X = Cl, Br, J) auf Ni$(C_5H_5)_2$ in quantitativer Ausbeute Salze der Zusammensetzung $[C_5H_5Ni(PBut_3)_2]X$ bilden [18].

Ganz ähnlich wie die Umsetzungen mit [PPh_3H]X verlaufen wahrscheinlich auch die Reaktionen von Dicyclopentadienylnickel mit Thiolen RSH (R = Me, Et, Ph) [19] und sekundären Phosphinen wie z.B. $P(CF_3)_2H$ [20], d.h. mit Verbindungen, die sowohl als Lewis-Base als auch als Protonendonator fungieren können [165]. In den nach (7) und (8) entstehenden Zweikernkomplexen sind die Metallatome jeweils durch SR- bzw. PR_2-Brücken verknüpft.

$$2\,Ni(C_5H_5)_2 + 2\,RSH \longrightarrow [C_5H_5NiSR]_2 + 2\,C_5H_6 \quad (7)$$

$$2\,Ni(C_5H_5)_2 + 4\,P(CF_3)_2H \longrightarrow [C_5H_5NiP(CF_3)_2]_2 + 2\,P(CF_3)_2C_5H_7 \quad (8)$$

Die Reaktion von Ni$(C_5H_5)_2$ mit C_3H_5MgCl, die bereits bei 0 °C in Tetrahydrofuran in etwa 50 %iger Ausbeute zu $C_5H_5NiC_3H_5$ führt [21], könnte in bezug auf ihren Verlauf der Umsetzung von Ni$(C_5H_5)_2$ mit PR_3 möglicherweise an die Seite gestellt werden. Ustynyuk und Mitarbeiter [14] nehmen an, daß gemäß (9) im Primärschritt durch nucleophilen Angriff des Allylanions am Nickel das Komplexanion

$$[(\pi\text{-}C_5H_5)Ni(\sigma\text{-}C_5H_5)C_3H_5]^- \text{ (analog zu } (\pi\text{-}C_5H_5)Ni(\sigma\text{-}C_5H_5)PR_3)$$

entsteht, das dann unter Spaltung der Ni-σ-C_5H_5-Bindung (π-C_5H_5)Ni(π-C_3H_5) ergibt. Der entsprechende π-Methallyl-Nickel-Komplex (π-C_5H_5)Ni(π-C_3H_4Me) ist im Sinne einer Ligandenverdrängung aus Ni$(C_5H_5)_2$

und Butadien zugänglich. Es wird dabei ein Isomerengemisch der syn- und anti-Form (I und II) erhalten [22].

$$Ni(C_5H_5)_2 + C_3H_5MgCl \longrightarrow [(\pi\text{-}C_5H_5)Ni(\sigma\text{-}C_5H_5)C_3H_5]^- MgCl^+$$
$$\longrightarrow (\pi\text{-}C_5H_5)Ni(\pi\text{-}C_3H_5) + C_5H_5MgCl \qquad (9)$$

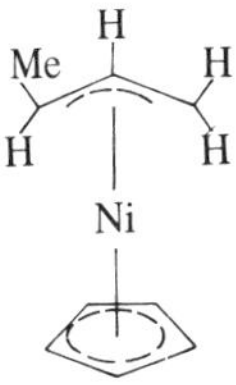

I

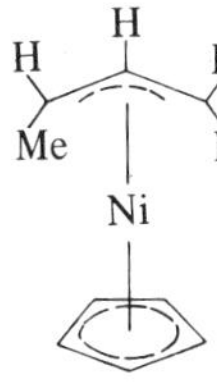

II

Das dem $C_5H_5NiC_3H_5$ formal isoelektronische C_5H_5NiNO – Prototyp und am längsten bekannter Vertreter der einkernigen Monocyclopentadienyl-Nickel-Verbindungen – bildet sich aus $Ni(C_5H_5)_2$ und NO, bei Verwendung von Petroläther bereits bei Raumtemperatur. Die Ausbeuten liegen bei ~ 50 % [23,24].

Zweikernkomplexe $[C_5H_5Ni]_2RC_2R'$ (III), deren Struktur derjenigen der oben beschriebenen Verbindungen $[C_5H_5NiSR]_2$ und $[C_5H_5NiP(CF_3)_2]_2$ (siehe (7) und (8)) und auch dem aus $Ni(C_5H_5)_2$ und $Ni(CO)_4$ zugänglichen $[C_5H_5NiCO]_2$ [25] sehr ähnlich ist, entstehen nach (10) bei der Reaktion von Dicyclopentadienylnickel und Alkinen RC_2R' ($R = R' = H$ [26]; $R = H$, $R' = CF_3$ [27]; $R = R' = CN$ [28]). Für die Darstellung von $[C_5H_5Ni]_2C_2Ph_2$ hat sich die zusätzliche Gegenwart von $Ni(CO)_4$ als nützlich erwiesen [29].

$2\,Ni(C_5H_5)_2 + RC_2R' \longrightarrow$ (10)

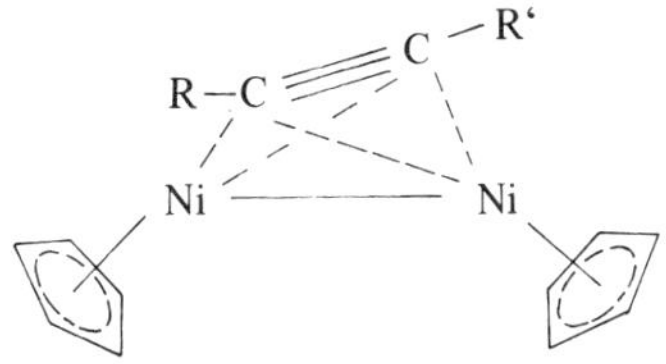

III

Wichtig für den Verlauf der Umsetzung von $Ni(C_5H_5)_2$ und RC_2R' scheint die Art der Alkinsubstituenten zu sein. Mit $C_2(COOMe)_2$ entsteht nicht ein zu III analoger Komplex sondern IV [30,31]. Auch olefinische π-Donoren reagieren mit $Ni(C_5H_5)_2$ bevorzugt unter Addition und nicht unter Substitution, d.h. Ligandenverdrängung [32].

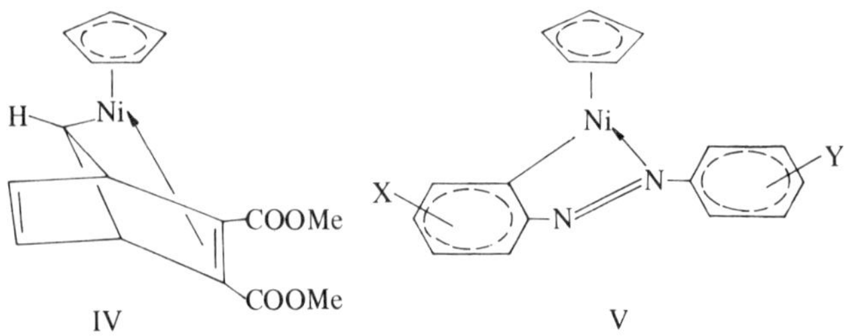

Eine zu (10) ähnliche Reaktion, die ebenfalls unter Spaltung einer π-C_5H_5-Ni-Bindung verläuft und zu Verbindungen des Typs V führt, erfolgt zwischen $Ni(C_5H_5)_2$ und Azobenzolen [15,33]. Da elektronenziehende Substituenten X oder Y eine Zunahme der Reaktionsgeschwindigkeit bewirken, ist die Vermutung geäußert worden, daß der Primärschritt in einem Einelektronentransfer und Bildung eines Ionenpaares von einem Nickelocinium-Kation und einem Azobenzolradikal-Anion besteht [15]. Auch für die Umsetzung von $Ni(C_5H_5)_2$ und CPh_3Cl, bei der gemäß (11) eine Spaltung *beider* π-C_5H_5-Ni-Bindungen stattfindet, ist als einleitender Schritt ein Einelektronenübergang anzunehmen [34].

$$Ni(C_5H_5)_2 + 2\,CPh_3Cl \longrightarrow NiCl_2 + 2\,C_5H_5CPh_3 \qquad (11)$$

Zu den Liganden-Verdrängungsreaktionen nach b) zählt schließlich noch die Umsetzung von $Ni(C_5H_5)_2$ und $S(N\text{-}t\text{-}C_4H_9)_2$ [35]. Es wäre denkbar, daß der dabei erhaltene Ni_3-Cluster VI ($R = t\text{-}C_4H_9$) nach (12) auf einem analogen Weg wie auch andere Monocyclopentadienyl-Nickel-Komplexe entstehen.

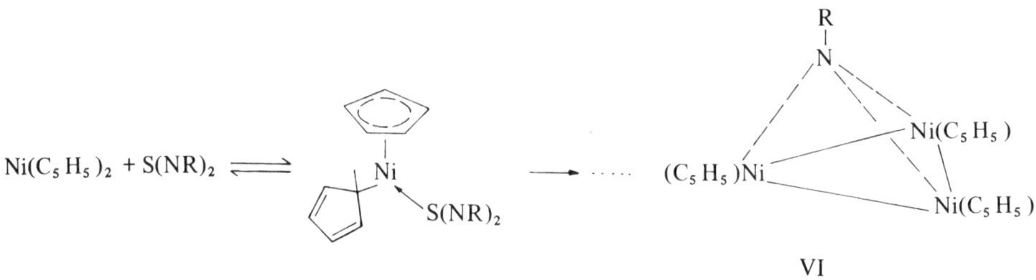

(12)

2.2. Reaktionen von Dicyclopentadienyleisen

Ferrocen $Fe(C_5H_5)_2$ verhält sich im Gegensatz zu $Ni(C_5H_5)_2$ gegenüber Lewis-Basen weitgehend inert. Erst in Gegenwart von Aluminiumtrihalogeniden AlX_3 (X = Cl, Br) gelingt mit aromatischen Sechsringliganden eine Ringaustauschreaktion [36,37]. Die formale Oxydationszahl des Eisens (+II) bleibt dabei erhalten.

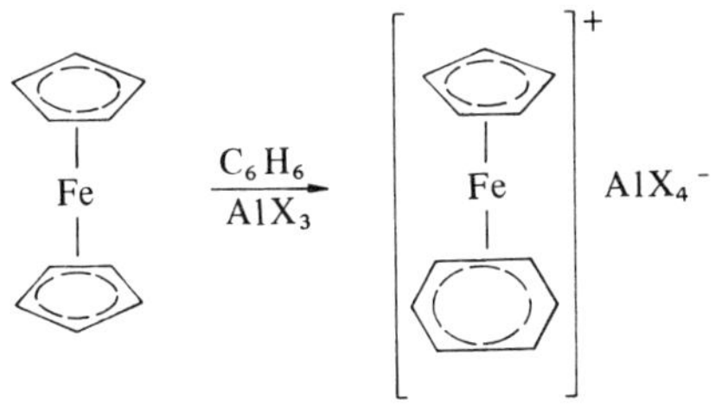

(13)

VII

Wie die eingehenden Untersuchungen von A. N. Nesmeyanov und Mitarbeitern gezeigt haben [38], hat es sich als am günstigsten erwiesen, einen 2–4fachen Überschuß an AlX_3 zu verwenden und die Umsetzung nach (13) bei der Siedetemperatur des jeweiligen Aromaten durchzuführen. Bei der Reaktion mit festen Liganden hat sich Decalin als Solvens oder ein Arbeiten in der Schmelze bewährt. In den Fällen, in denen an den Fünfringen oder am Sechsring nicht ein leicht reduzierbarer Substituent wie z.B. Cl, Br, CN oder COMe vorhanden ist, empfiehlt es sich, dem Reaktionsgemisch Aluminiumpulver (im Mengenverhältnis $Fe(C_5H_5)_2$: Al = 1 : 1) zuzusetzen; dadurch wird eine Oxydation des Ferrocens durch AlX_3 zum Ferrocinium-Kation $[Fe(C_5H_5)_2]^+$ verhindert. Ein Ringaustausch an diesem kationischen Komplex – analog zu (13) – ist bisher noch nicht gelungen.

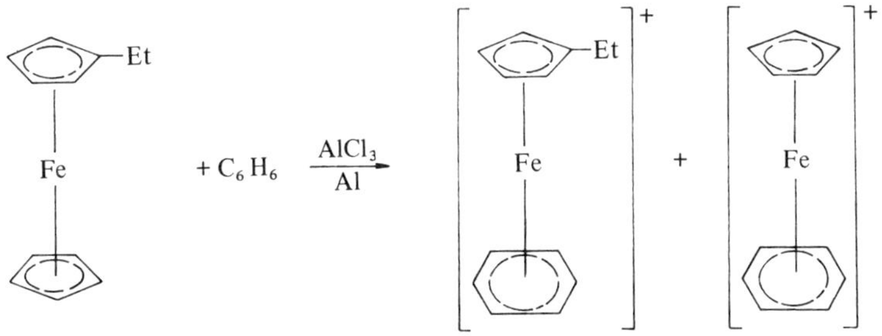

(14)

(27%) (73%)

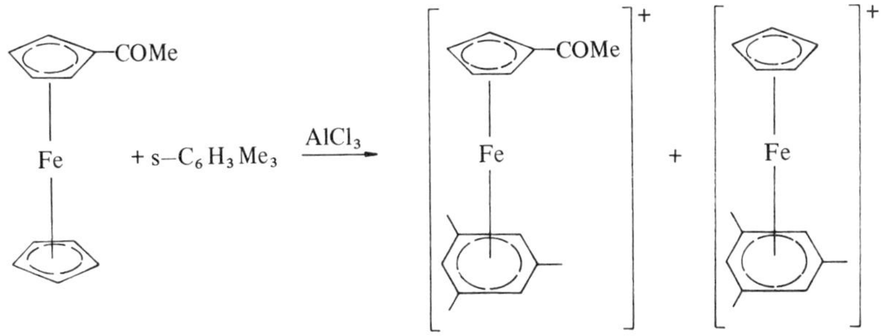

(15)

(80–90%) (10–20%)

Neben C_6H_6 reagieren auch zahlreiche substituierte Benzolderivate sowie annellierte Ringe wie Naphthalin, Tetralin und Fluoren mit $Fe(C_5H_5)_2$ zu den entsprechenden Aromaten-eisen-cyclopentadienyl-Kationen. Ebenso sind Verbindungen $[RC_5H_4FeC_6R'_6]X$ ausgehend von substituierten Ferrocenen und Aromaten $C_6R'_6$ zugänglich [36-41]. Die Substituenten R am abgespaltenen Cyclopentadienylring und R′ am aromatischen Liganden sind maßgebend für die Reaktionsgeschwindigkeit: Elektronenziehende Gruppen erschweren, elektronenliefernde Gruppen erleichtern den Ringaustausch. Daher reagiert 1.1′-Diäthylferrocen schneller, 1.1′-Diacetylferrocen jedoch wesentlich langsamer als $Fe(C_5H_5)_2$. Die Ausbeuten der nach (14) und (15) erhaltenen Reaktionsprodukte unterstreichen ebenfalls die Bedeutung des Substituenteneinflusses [38].

Eng verwandt mit den Ligandenverdrängungsreaktionen nach (13) – (15) sind die sogenannten „Synproportionierungs-“ oder Ligandenübertragungsreaktionen bei denen z.B. ausgehend von monosubstituierten Ferrocenen $C_5H_4RFeC_5H_5$ bei Anwesenheit von $AlCl_3$ oder $LiAlH_4$ zu gleichen Teilen die disubstituierten und unsubstituierten Verbindungen $Fe(C_5H_4R)_2$ und $Fe(C_5H_5)_2$ entstehen [38,42]. Umsetzungen dieser Art werden zusammenfassend unter 4. besprochen.

Erwähnt seien hier auch noch die Reaktionen der Kationen $[Fe(C_5H_5)_2]^+$ und $[C_5H_5FeC_6H_6]^+$ mit β-Diketonen $RCOCH_2COR$ (R = Me, Ph), die zu den Komplexen $C_5H_5Fe(RCOCHCOR)_2$ führen [43,44]. Die Spaltung der Eisen-Sechsring-Bindung in $[C_5H_5FeC_6H_6]^+$ erfolgt bei niedrigeren Temperaturen als die Spaltung der Eisen-Fünfring-Bindung in $[Fe(C_5H_5)_2]^+$. Es ist vermutet worden, daß sich bei der Umsetzung des kationischen Cyclopentadienyl-Benzol-Komplexes mit $RCOCH_2COR$ primär eine höher koordinierte Zwischenverbindung bildet, die dann unter Abgabe von C_6H_6, d.h. des Liganden mit der kleineren Bindungsenergie, in das Endprodukt übergeht [44].

2.3. Reaktionen von Komplexen $M(C_5H_5)_n$ ($M \neq Fe, Ni$)

Über Ligandenverdrängungsreaktionen anderer *reiner* Metallcyclopentadienyl-Komplexe als $Fe(C_5H_5)_2$ und $Ni(C_5H_5)_2$ ist relativ wenig bekannt. Auf Grund der bisher erhaltenen Ergebnisse dürfte es jedoch durchaus aussichtsreich sein, vor allem Verbindungen wie z.B. $Co(C_5H_5)_2$ oder $Mn(C_5H_5)_2$ als Ausgangssubstanzen für die Darstellung von Kobalt- und Mangan-Komplexen niederer Oxydationsstufen zu verwenden.

Am längsten bekannt von Umsetzungen zwischen $M(C_5H_5)_n$ und L sind solche mit der Lewis-Base Kohlenmonoxyd, die entweder zu Cyclopentadienylmetall-carbonylen oder zu reinen Metallcarbonylen, in Gegenwart von H_2 auch zu Cyclopentadienyl-metall-carbonyl-hydriden führen können. Der Einfluß der Reaktionsbedingungen auf die Zusammensetzung der entstehenden Produkte ist sehr treffend am Beispiel der Umsetzung von $Cr(C_5H_5)_2$ mit CO zu erkennen:

Bei 100–200 at CO und 100–110 °C entsteht gemäß (16) der salzartige Komplex $[Cr(C_5H_5)_2]^+[C_5H_5Cr(CO)_3]^-$.

$$2\,Cr(C_5H_5)_2 + 3\,CO \longrightarrow [Cr(C_5H_5)_2][C_5H_5Cr(CO)_3] + [C_5H_5\cdot] \qquad (16)$$

Bei Steigerung der Temperatur auf 150–170 °C bildet sich nach (17) die CO-verbrückte, zweikernige Verbindung $[C_5H_5Cr(CO)_3]_2$ und unter noch schärferen Bedingungen bei ~ 250 °C findet nach (18) eine vollständige Verdrängung der Cyclopentadienyl-Liganden unter Bildung von $Cr(CO)_6$ statt [45].

$$2\,Cr(C_5H_5)_2 + 6\,CO \longrightarrow [C_5H_5Cr(CO)_3]_2 + C_{10}H_{10} \qquad (17)$$

$$Cr(C_5H_5)_2 + 6\,CO \longrightarrow Cr(CO)_6 + C_{10}H_{10} \qquad (18)$$

Bei der Reaktion von $Cr(C_5H_5)_2$ mit einem CO/H_2-Gemisch und Einhaltung einer Temperatur von ~ 70 °C wird der Hydrid-Komplex $C_5H_5Cr(CO)_3H$ erhalten [45,46]. Fischer und Hafner haben die Vermutung geäußert, daß die Ligandenverdrängung in jedem Fall über die Bildung von $C_5H_5Cr(CO)_3$-Radikalen verläuft, die entweder mit $Cr(C_5H_5)_2$ im Sinn eines Redoxprozesses, mit CO unter weiterer Substitution, mit H_2 unter Spaltung der Wasserstoffmolekel, oder aber unter Dimerisierung reagieren. In Analogie zu den Reaktionen von $Ni(C_5H_5)_2$ mit Lewis-Basen wäre es andererseits auch denkbar, daß intermediär

Tabelle 2. *Reaktionen von Metallcyclopentadienyl-Komplexen $M(C_5H_5)_n$ mit CO, CO/H_2 und NO*

Ausgangsverbindung	Reaktionsbedingungen	Produkt	Ausbeute (in %)	Ref.
$Co(C_5H_5)_2$	200 at CO; 90–150 °C	$C_5H_5Co(CO)_2$		47)
$Mn(C_5H_5)_2$	200–250 at CO; 90–150 °C	$C_5H_5Mn(CO)_3$	80	48,49)
$Mn(C_5H_5)_2$	55 at CO; 250 °C	$C_5H_5Mn(CO)_3$	40	24)
$Mn(C_5H_4Me)_2$	70 at CO; 200 °C	$C_5H_4MeMn(CO)_3$	70	50)
$Mn(C_9H_7)_2$	210 at CO; 150 °C; THF	$C_9H_7Mn(CO)_3$		51)
$Re(C_5H_5)_2H$	50 at CO; 225 °C; Dimethylglykoläther	$C_5H_5Re(CO)_3$	16	52)
$Re(C_5H_5)_2H$	250 at CO; 90 °C	$C_5H_5ReC_5H_6(CO)_2$	55	53,54)
$Cr(C_5H_5)_2$	150–200 at CO; 110 °C	$[Cr(C_5H_5)_2][C_5H_5Cr(CO)_3]$	50	45)
$Cr(C_5H_5)_2$	100 at CO; 150–170 °C	$[C_5H_5Cr(CO)_3]_2$		45)
$Cr(C_5H_5)_2$	150 at CO; 50 at H_2; 70 °C	$C_5H_5Cr(CO)_3H$	45	45,46)
$V(C_5H_5)_2$	200–250 at CO; 70–80 at H_2; 140 °C	$C_5H_5V(CO)_4$	97	55,56)
$Ti(C_5H_5)_3$	150 at CO; 80 °C	$(C_5H_5)_2Ti(CO)_2$		57)
$Ti(C_5H_5)_n$	135 at CO; 100 °C	$(C_5H_5)_2Ti(CO)_2$	18	58)
$Mn(C_5H_5)_2$	NO; 25 °C; THF	$(C_5H_5)_3Mn_2(NO)_3$	30	59,60)

eine π-Cyclopentadienyl-σ-cyclopentadienyl-Chrom-Verbindung wie z.B. $(\pi\text{-}C_5H_5)Cr(CO)_3(\sigma\text{-}C_5H_5)$ entsteht, aus der durch homolytische oder heterolytische Spaltung der Cr-σ-C_5H_5-Bindung je nach den Reaktionsbedingungen sich die verschiedenen Produkte bilden.

Zusammenfassende Angaben über Umsetzungen von Metallcyclopentadienyl-Komplexen $M(C_5H_5)_n$ mit CO, CO/H_2 und NO, über Reaktionsbedingungen, Zusammensetzung der Reaktionsprodukte und Ausbeuten sind aus Tabelle 2 zu entnehmen.

Ligandenverdrängungsreaktionen von $Co(C_5H_5)_2$ sind in letzter Zeit auch auf die Koordinationspartner PF_3 [61] und $P(OR)_3$ (R = Me, Et, Ph) [62] ausgedehnt worden. Während die Darstellung von $C_5H_5Co(PF_3)_2$ im Autoklaven bei 170 °C und einem PF_3-Druck von 300 at gelingt, verlaufen die analogen Umsetzungen mit tertiären Phosphiten gemäß (19) in siedendem Dioxan mit sehr guten Ausbeuten. Für R = Me und Et entstehen neben den monomeren Komplexen $C_5H_5Co[P(OR)_3]_2$ noch thermisch sehr stabile, höhermolekulare Reaktionsprodukte, die wahrscheinlich einen neuen Typ sandwichartig gebauter Metall-Cluster-Verbindungen darstellen [12].

$$Co(C_5H_5)_2 + 2\,P(OR)_3 \longrightarrow C_5H_5Co[P(OR)_3]_2 + [C_5H_5\cdot] \qquad (19)$$

Aus $Co(C_5H_5)_2$ und Azobenzol ist bei ~ 130 °C in sehr geringer Ausbeute ein

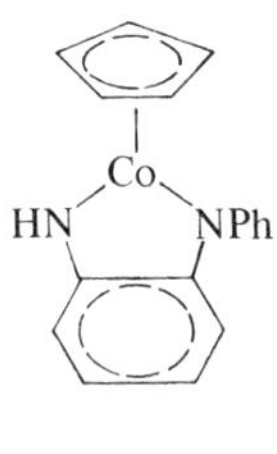

VIII

Komplex VIII zugänglich, in dem nach Aussage vorläufiger Strukturuntersuchungen das Diamid des o-Aminodiphenylamins als Chelatligand gebunden vorliegt [63].

2.4. Reaktionen von gemischten Cyclopentadienyl-Metall-Komplexen

Reaktionen von Cyclopentadienyl-Metall-Komplexen des Typs $(C_5H_5)_nML'_p$ bzw. $(C_5H_5)_nML'_pX_q$ mit Lewis-Basen (L) sind schon sehr zahlreich studiert worden; bei den meisten von ihnen erfolgt allerdings ein Austausch von L' oder X gegen L und nicht eine Ligandenverdrängung unter Spaltung der Ring-Metall-Bindung.

Auf die Verwendung der Verbindungen $(C_5H_5)_nML'_p$ und $(C_5H_5)_nML'_pX_q$ für die Synthese vor allem von *Metall(0)-Komplexen* haben gerade neuere Arbeiten von H. Behrens und Mitarbeitern aufmerksam gemacht. So gelingt ausgehend von $C_5H_5V(CO)_4$ und den mehrzähnigen Liganden dipy, tripy und phen unter vollständiger Verdrängung des Fünfrings und der CO-Gruppen die Darstellung von $V(dipy)_3$, $V(tripy)_2$ und $V(phen)_3$. Reaktionstemperaturen von 120–150 °C und Cyclohexan als Lösungsmittel haben sich dabei als günstig erwiesen [64,65]. $[C_5H_5Fe(CO)_2]_2$ reagiert mit tripy bei 160 °C in Benzol zu $Fe(CO)_2$tripy, eine der wenigen bis heute bekannten Eisendicarbonyl-Verbindungen [66]. Bei der Umsetzung von $[C_5H_5Fe(CO)_2]_2$ mit 1.2-Bis(diphenylphosphino)äthan in Cyclohexan gemäß (20) erfolgt formal eine Art „Valenzdisproportionierung", und zwar unter gleichzeitiger Wanderung eines Cyclopentadienyl-Liganden [66].

$$[C_5H_5Fe^{+I}(CO)_2]_2 + C_2H_4(PPh_2)_2 \longrightarrow Fe^{+II}(C_5H_5)_2 + Fe^0(CO)_3[C_2H_4(PPh_2)_2] + CO \qquad (20)$$

Ähnliche Disproportionierungsvorgänge (siehe (21) bis (24)) werden auch bei den Umsetzungen von flüssigem NH_3 oder Lösungen von KCN in flüssigem NH_3 mit $[C_5H_5Fe(CO)_2]_2$ und $[C_5H_5NiCO]_2$ [67,166] sowie bei der Einwirkung von wasserfreiem Hydrazin auf $C_5H_5Co(CO)_2$ [68] beobachtet; die Beteiligung des Hydrazins an der Reaktion (24) ist wahrscheinlich im Sinne einer Basenkatalyse zu verstehen.

$$2\,[C_5H_5Fe^{+I}(CO)_2]_2 + 6\,NH_3 \longrightarrow 2\,Fe^{+II}(C_5H_5)_2 + [Fe^{+II}(NH_3)_6][Fe^{-II}(CO)_4] + 4\,CO \qquad (21)$$

$$[C_5H_5Ni^{+I}CO]_2 + 8\,NH_3 \longrightarrow [Ni^{+II}(NH_3)_6](C_5H_5)_2 + Ni^0(CO)_2(NH_3)_2 \qquad (22)$$

$$[C_5H_5Ni^{+I}CO]_2 + 6\,KCN \longrightarrow K_2[Ni^{+II}(CN)_4] + K_2[Ni^0(CO)_2(CN)_2] + 2\,KC_5H_5 \qquad (23)$$

$$2\,C_5H_5Co^{+I}(CO)_2 \longrightarrow [Co^{+III}(C_5H_5)_2][Co^{-I}(CO)_4] \qquad (24)$$

Der gemischte Carbonyl-Nitrosyl-Komplex $C_5H_5Mo(CO)_2NO$ reagiert mit KCN in flüssigem NH_3 gemäß (25) zu $K_4[Mo(CN)_5NO]$, d.h. es erfolgt eine Verdrängung des C_5H_5- und der zwei CO-Liganden unter Erhalt der formalen Oxydationszahl Null des Metallatoms [69].

$$C_5H_5Mo^0(CO)_2NO + 5\,KCN \longrightarrow K_4[Mo^0(CN)_5NO] + 2\,CO + KC_5H_5 \qquad (25)$$

Neutralliganden wie z.B. dipy, phen, PPh_3, $AsPh_3$ oder $SbPh_3$ bewirken bei der Umsetzung mit $C_5H_5Mo(CO)_2NO$ auch beim Erhitzen bis auf etwa 250 °C keine Substitution [69]. Die Reaktion von $[C_5H_5FeNO]_2$ mit PPh_3 führt in Xylol unter Rückfluß zur Bildung von $Fe(NO)_2(PPh_3)_2$ [70]. Hierbei findet neben der Verdrängung des Cyclopentadienylrings offensichtlich auch eine intermolekulare Wanderung einer NO-Gruppe statt.

Als Nebenreaktion wird die Spaltung der Ring-Metall-Bindung bei der Einwirkung von $PBut_3$ und $PBut_2Ph$ auf $C_5H_5Mo(CO)_3Cl$ und $C_9H_7Mo(CO)_3X$ (X = Br, J) beobachtet [71,72]. Die Hauptprodukte sind hierbei die Verbindungen

$$C_5H_5Mo(CO)_2LCl \text{ bzw. } C_9H_7Mo(CO)_2LX;$$

daneben entstehen in geringerer Ausbeute die Carbonyl-Phosphin-Komplexe $Mo(CO)_3L_3$ und $Mo(CO)_4L_2$ (L = $PBut_3$, $PBut_2Ph$). Ihre Bildung kann nach (26) und (27) formuliert werden (R = C_5H_5, C_9H_7).

$$RMo(CO)_3X + 4\,L \longrightarrow Mo(CO)_3L_3 + [RL]^+ + Cl^- \tag{26}$$

$$Mo(CO)_3L_3 + CO \longrightarrow Mo(CO)_4L_2 + L \tag{27}$$

Bei den entsprechenden Umsetzungen mit weniger basischen Phosphinen werden nur die Verbindungen $C_5H_5Mo(CO)_2LCl$ bzw. $C_9H_7Mo(CO)_2LX$ erhalten [71,72].

King und Mitarbeiter [73] haben kürzlich über eine neue, bei Normaldruck durchführbare Synthese von $Mn_2(CO)_{10}$ berichtet, die sich einer Ringliganden-Verdrängungsreaktion bedient. Als Ausgangsmaterial dient dabei $C_5H_4MeMn(CO)_3$, das in Diglyme mit Natrium und CO (1 at) umgesetzt wird. Die Ausbeute ist mit 15–20 % zwar niedriger als bei der entsprechenden Reaktion unter höherem CO-Druck [74], doch wird dafür das Arbeiten im Autoklaven vermieden. Als Zwischenverbindungen entstehen bei dieser Synthese möglicherweise radikalische Species wie z.B. $Mn(CO)_3$(diglyme), die dann weiter mit Kohlenoxid reagieren. Die Metall-Cyclopentadienyl-Bindung in $C_5H_5Mn(CO)_3$ und $C_5H_4MeMn(CO)_3$ ist im allgemeinen sehr stabil; unter „normalen" Bedingungen erfolgt bei der Umsetzung mit L keine Ringligandenverdrängung [75].

Eine ziemlich breite Anwendung haben Verbindungen des Typs $(C_5H_5)_nML'_p$ oder $(C_5H_5)_nML'_pX_q$ als Ausgangssubstanzen für die Darstellung von Komplexen mit 1.2-Dithiolato- und Dithiocarbamato-Liganden gefunden. Bei den entsprechenden Reaktionen tritt vor allem dann die Spaltung einer Metall-Ring-Bindung ein, wenn in Gegenwart von N_2H_4 oder OR^- gearbeitet wird. Die Rolle dieser Agenzien ist in den meisten Fällen nicht die eines Reduktionsmittels, da bei der erfolgenden Ligandenverdrängung die formale Oxydationszahl des Metalls erhalten bleibt.

Die bisher bekannten Ergebnisse können wie folgt zusammengefaßt werden:

a) Bei den Umsetzungen von $C_5H_5Mo(CO)_2NO$, $[C_5H_5Mo(NO)J_2]_2$ und $C_5H_5Mo(CO)_3J$ mit überschüssigem $S_2C_2(CF_3)_2$, $[S_2C_2(CN)_2]^{2-}$oder $[S_2C_6Cl_4]^{2-}$ entstehen als Hauptprodukte Anionen des Typs $[Mo(S_2C_2R_2)_3]^{2-}$[76-78]. Die Einwirkung von $[Me_2NC(S)S]_2$ auf $C_5H_5Mo(CO)_2NO$ führt in geringen Ausbeuten zur Bildung des heptakoordinierten, ungeladenen Komplexes $[Mo(NO)(S_2CNMe_2)_3]$[76].

b) $[C_5H_5Mn(CO)_2NO]PF_6$ reagiert mit $[S_2C_6Cl_4]^{2-}$ in Gegenwart von KOEt in EtOH zu $[Mn(NO)(S_2C_6Cl_4)_2]^{2-}$[76]. Das analoge Anion $[Mn(NO)(S_2C_2(CF_3)_2)_2]^{2-}$ wird ausgehend von $C_5H_5Mn(NO)S_2C_2(CF_3)_2$ und N_2H_4 in Aethanol erhalten. Hierbei findet gleichzeitig zu der Spaltung der C_5H_5-Mn-Bindung eine intermolekulare Übertragung eines Dithiolat-Liganden statt [79].

c) Die Reaktion von $[C_5H_5FeNO]_2$ mit $[S_2C_2(CN)_2]^{2-}$oder $[S_2CNMe_2]^-$ergibt bereits unter sehr schonenden Bedingungen die Komplexe $[Fe(NO)(S_2C_2(CN)_2)_2]^{2-}$und $[Fe(NO)(S_2CNMe_2)_2]$[76].

d) Bei der Einwirkung von überschüssigem $[S_2C_2(CN)_2]^{2-}$auf $C_5H_5Co(CO)J_2$ in Aceton/Äthanol bildet sich – vermutlich über die Zwischenverbindung $C_5H_5CoS_2C_2(CN)_2$ - das Anion $[Co(S_2C_2(CN)_2)_3]^{3-}$, d.h. ein Komplex des dreiwertigen Kobalts [77]. Dithiolato-Komplexe von Co(II) der Zusammensetzung $[Co(S_2C_2R_2)_2]^{2-}$($R = CF_3$, CN) entstehen ausgehend von $C_5H_5CoS_2C_2R_2$ und $[S_2C_2R_2]^{2-}$ oder N_2H_4/EtOH [76,79]. Analog sind auch die Nickel(II)-Verbindungen $[Ni(S_2C_2R_2)_2]^{2-}$, und zwar durch Umsetzung von $[C_5H_5NiCO]_2$ (oder $Ni(C_5H_5)_2$ [77])mit $[S_2C_2(CN)_2]^{2-}$ [76] oder durch Disproportionierung von $C_5H_5NiS_2C_2(CF_3)_2$ in Gegenwart von N_2H_4/EtOH [79] zugänglich.

e) Die Spaltung einer Cyclopentadienyl-Metall-Bindung erfolgt auch bei den Umsetzungen von 1 Mol $(C_5H_5)_2TiCl_2$ mit 2 Mol $[S_2C_2(CN)_2]^{2-}$[77] oder $[S_2C_6Cl_4]^{2-}$ [76], die zu den Anionen $[C_5H_5Ti(S_2C_2R_2)_2]^-$ führen. Die dabei primär entstehenden Zwischenverbindungen $(C_5H_5)_2TiS_2C_2R_2$ können bei Wahl eines Molverhältnisses von $(C_5H_5)_2TiCl_2 : [S_2C_2R_2]^{2-} = 1:1$ gefaßt werden. Längeres Erhitzen von $(C_5H_5)_2TiS_2C_6Cl_4$ oder $[C_5H_5Ti(S_2C_6Cl_4)_2]^-$ mit überschüssigem $[S_2C_6Cl_4]^{2-}$in Aceton/Äthanol ergibt $[Ti(S_2C_6Cl_4)_3]^{2-}$[76].

Der Reaktion von $(C_5H_5)_2TiCl_2$ mit $[S_2C_2R_2]^{2-}$ verwandt ist die Umsetzung von $(C_5H_5)_2ZrCl_2$ mit Diketonen, die gemäß (28) unter Substitution der Cl-Liganden und Verdrängung eines Cyclopentadienylrings in sehr guter Ausbeute zu den Komplexen $[C_5H_5Zr(RCOCHCOR')_3]$ IX ($R = CF_3$; $R' = H, CF_3$) führt [80].

$$(C_5H_5)_2ZrCl_2 + 3\ RCOCH_2COR' \longrightarrow C_5H_5Zr(RCOCHCOR')_3 + C_5H_6 + 2\ HCl \qquad (28)$$

Die entsprechende Verbindung mit $R = R' = t\text{-}C_4H_9$ ist durch Umsetzung von $Zr(C_5H_5)_4$ und Dipivaloylmethan in siedendem Dichloräthan erhältlich [80].

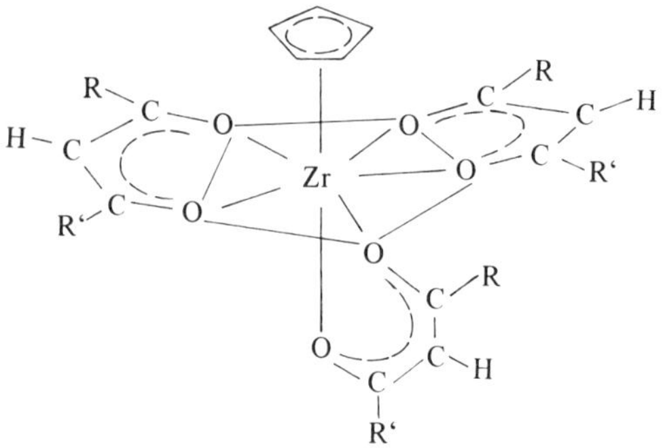

IX

2.5. Reaktionen von Dibenzol-Metall-Komplexen

Eine Verdrängung der Sechsringliganden in den ungeladenen Dibenzol-Metall-Komplexen $M(C_6R_6)_2$ (von denen bis heute Vertreter von Metallen der 5., 6. und 8. Nebengruppe bekannt sind) ist bisher nur in wenigen Fällen gelungen. Nach den vorliegenden Ergebnissen zu schließen, bedarf es für die Spaltung der Ring-Metall-Bindung ziemlich drastischer Reaktionsbedingungen.

Der am leichtesten zugängliche und in seinen Eigenschaften wohl am besten untersuchte Komplex, das *Dibenzolchrom,* reagiert mit CO [81] oder mit PF_3 [82] bei hohem Druck (320 at CO; 350 at PF_3) und ~ 200 °C zu $Cr(CO)_6$ bzw. $Cr(PF_3)_6$. Für die entsprechende Darstellung von $Mo(PF_3)_6$ aus $Mo(C_6H_6)_2$ hat sich ein PF_3-Druck von 600 at und eine Temperatur von 100 °C als günstig erwiesen [83]. Bei geringerem PF_3-Druck und höherer Temperatur wird hauptsächlich Zersetzung beobachtet. Die *Molybdän-Verbindungen* Mo(diphosphin)$_3$ sind durch Umsetzung von $Mo(C_6H_6)_2$ mit den Diphosphinen 1.2-$C_2H_4(PMe_2)_2$, 1.2-$C_2H_4(PPh_2)_2$ und o-$C_6H_4(PEt_2)_2$ im geschlossenen System bei 140 °C zugänglich [84]. $Cr(C_6H_6)_2$ reagiert auch beim Erhitzen bis auf 200 °C nicht mit den angegebenen Diphosphinen.

Die Reaktionen mehrzähniger Stickstoff-Donoren, wie z.B. dipy, tripy und phen, mit $Cr(C_6H_6)_2$ oder $Mo(C_6H_6)_2$ führen nur mit tripy zu einer Ligandenverdrängung. Im Einschlußrohr bei 160 °C und in Cyclohexan als Solvens entstehen gemäß (29) die Komplexe M(tripy)$_2$ [85].

$$M(C_6H_6)_2 + 2\ \text{tripy} \longrightarrow M(\text{tripy})_2 + 2\ C_6H_6 \qquad (29)$$

Das paramagnetische *Dibenzolvanadin* setzt sich – ebenso wie $V(C_5H_5)_2$ und $C_5H_5V(CO)_4$ – mit tripy in Cyclohexan bei 120 °C quantitativ zu V(tripy)$_2$ um [64].

Bei der Einwirkung von CO auf $V(s\text{-}C_6H_3Me_3)_2$ findet neben einer Metall-Ring-Spaltung gleichzeitig auch ein Redoxvorgang statt. Nach Untersuchungen von Calderazzo und Cini [86] bildet sich dabei gemäß (30) der salzartige Komplex $[V(s\text{-}C_6H_3Me_3)_2][V(CO)_6]$, der auch ausgehend von $V(s\text{-}C_6H_3Me_3)_2$ und $V(CO)_6$ zugänglich ist.

$$2\,V(s\text{-}C_6H_3Me_3)_2 + 6\,CO \longrightarrow [V(s\text{-}C_6H_3Me_3)_2][V(CO)_6] + 2\,s\text{-}C_6H_3Me_3 \qquad (30)$$

Aus dem Diaromaten-Komplex und CO entsteht wahrscheinlich primär (möglicherweise über die Zwischenverbindung $s\text{-}C_6H_3Me_3V(CO)_3)V(CO)_6$, das in einem rasch verlaufenden Folgeschritt von überschüssigem $V(s\text{-}C_6H_3Me_3)_2$ zum $[V(CO)_6]^-$-Anion reduziert wird.

Während – wie oben erwähnt – das ungeladene $Cr(C_6H_6)_2$ gegenüber Lewis-Basen außerordentlich inert ist, läßt sich an dem bei der Dibenzolchrom-Synthese nach Fischer-Hafner zunächst entstehenden „Primärkomplex" der Zusammensetzung $3[Cr(C_6H_6)_2]AlCl_4 \cdot 4\,AlCl_3$ ein Ligandenaustausch bereitwillig realisieren: Erhitzen des „Primärkomplexes" mit Mesitylen liefert quantitativ $[Cr(s\text{-}C_6H_3Me_3)_2]^+$; eine Rückreaktion zu $[Cr(C_6H_6)_2]^+$ erfolgt in Gegenwart von $AlCl_3$ und überschüssigem Benzol [87,88]. Analoge, durch $AlCl_3$-katalysierte Austauschreaktionen von Sechsringliganden wurden von Hein und Kartte [89] auch am $[Cr(C_6H_6)(C_6H_5C_6H_5)]^+$ und $[Cr(C_6H_5C_6H_5)_2]^+$ untersucht und zur Darstellung von zahlreichen Kationen $[Cr(Aromat)_2]^+$ präparativ genutzt. Ausgehend von dem nach der Grignard-Methode leicht zugänglichen $[Cr(C_6H_6)(C_6H_5C_6H_5)]^+$ entsteht z.B. durch zweistündiges Erhitzen in Benzol mit etwa 50 %iger Ausbeute $[Cr(C_6H_6)_2]^+$ [89].

$$[Cr(C_6H_6)(C_6H_5C_6H_5)]^+ + C_6H_6 \xrightarrow{AlCl_3} [Cr(C_6H_6)_2]^+ + C_6H_5C_6H_5 \qquad (31)$$

Bemerkenswert erscheint der Hinweis [89], daß bei Ersatz von Benzol durch dipy oder PPh_3 in der nach (31) formulierten Reaktion kein Ligandenaustausch zu beobachten ist. Die starke Lewis-Säure $AlCl_3$ blockiert wahrscheinlich vollständig die Donoreigenschaften der Liganden dipy und PPh_3, so daß diese nicht mehr mit dem Metallatom in Wechselwirkung treten können.

Die *Verdrängung eines Sechsrings* unter dem Einfluß von $AlCl_3$ gelingt ebenfalls bei den Umsetzungen von $C_5H_5CrC_6H_6$ und 1.3.5-Cycloheptatrienen C_7H_7R (R = H, Me, Ph) gemäß (32) [90].

$$C_5H_5CrC_6H_6 + C_7H_7R \xrightarrow{AlCl_3} [C_5H_5CrC_7H_6R]^+ + C_6H_6 + H^- \qquad (32)$$

Bei Verwendung von $[C_7H_7]BF_4$ anstatt von C_7H_8 erfolgt ein Austausch auch in Abwesenheit von $AlCl_3$; die Ausbeute an $[C_5H_5CrC_7H_7]^+$ ist dann allerdings wesentlich geringer als nach der $C_7H_7R/AlCl_3$-Methode [90]. Die Reaktion von $C_5H_5CrC_6H_6$ mit Azulen und $BF_3 \cdot OMe_2$ in Benzol führt zur Bildung eines dem $[C_5H_5CrC_7H_7]^+$ analogen Azulenium-chrom-cyclopentadienyl-Kations $[C_5H_5CrC_{10}H_9]^+$ [91]. Bei all diesen Verdrängungsreaktionen findet im Primärschritt vermutlich eine Wechselwirkung zwischen der Lewis-Säure ($AlCl_3$, BF_3 oder $[C_7H_7]^+$) und dem Sechsring statt; dadurch wird die Chrom-Benzol-Bindung geschwächt und der Austausch gegen einen anderen Ringliganden erleichtert.

2.6. Reaktionen von Benzol-Metalltricarbonyl-Komplexen

Die Halbsandwich-Komplexe $C_6R_6M(CO)_3$ (M = Cr, Mo, W) reagieren mit Lewis-Basen wesentlich bereitwilliger als die Sandwich-Verbindungen $M(C_6R_6)_2$. Bei Zufuhr thermischer Energie erfolgt bei der Umsetzung von $C_6R_6M(CO)_3$ und L praktisch ohne Ausnahme eine Spaltung der Ring-Metall-Bindung unter Bildung von cis-$M(CO)_3L_3$ (L = einzähniger Ligand). Bei analogen photochemischen Reaktionen dominiert die Substitution einer CO-Gruppe; es entstehen dann Verbindungen $C_6R_6M(CO)_2L$.

Beispiele für Umsetzungen gemäß (33) sind für

M = Cr, Mo, W und

L =	Lit.	L =	Lit.
PF_3	[82,83]	$P(OMe)_2Me$	[98]
PCl_3	[92,93]	$AsMe_2Ph$	[94]
PCl_2Ph	[93]	NH_3	[85]
$P(Alkyl)_3$	[92,93]	Pyridin	[94]
PPh_3	[92,94]	Piperazin	[99]
$P(OMe)_3$	[92,95,96]	Morpholin	[99]
$P(OPh)_3$	[97]		

bekannt.

$$C_6R_6M(CO)_3 + 3\,L \longrightarrow \text{cis-}M(CO)_3L_3 + C_6R_6 \qquad (33)$$

Für M = Mo sind Temperaturen von 20–40 °C, für M = Cr und W solche von 60–150 °C üblich. Als Lösungsmittel wurden Benzol, Hexan, Heptan, Chloroform, 1.2-Dichloräthan – verschiedentlich auch der reine Ligand – verwendet. Die Ausbeute ist in den meisten Fällen quantitativ.

Mit mehrzähnigen Liganden tritt ebenfalls sehr leicht Reaktion ein: Die Umsetzung von Mesitylen-molybdän-tricarbonyl mit Diäthylentriamin (dien) führt z.B. bei Raumtemperatur in Äther zu $(dien)Mo(CO)_3$ [100]. Mit dem Ketazin X reagiert s-$C_6H_3Me_3Mo(CO)_3$ unter gleichen Bedingungen zu dem Komplex XI, in dem die beiden N-N-Liganden einzähnig koordiniert sind [101]. Eine entsprechende Verbindung cis-$Mo(CO)_4L_2$ entsteht auch aus s-$C_6H_3Me_3Mo(CO)_3$ und 2.3-Dimethyl-5.6-dihydropyrazin.

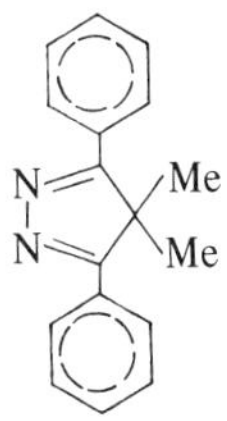

X

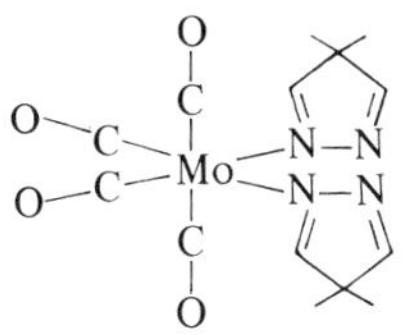

XI

Bei wesentlich höheren Temperaturen (160–220 °C) ist bei den Umsetzungen von dipy, tripy und phen mit $C_6H_6Cr(CO)_3$ auch eine *Totalsubstitution* der C_6H_6-*und* CO-Gruppen möglich; es entstehen dann die Verbindungen $Cr(dipy)_3$, $Cr(tripy)_2$ und $Cr(phen)_3$ [85]. In diesem Zusammenhang sei daran erinnert, daß ausgehend von $Cr(C_6H_6)_2$ und den gleichen Chelatbildnern nur mit tripy ein vollständiger Ligandenaustausch gelingt.

Zwei weitere Reaktionen von Aromaten-chrom-tricarbonylen, bei denen sowohl eine Verdrängung des Sechsrings als auch der CO-Gruppen (in dem einen Fall unter gleichzeitiger Oxydation von Cr(0) zum Cr(III)) eintritt, seien hier kurz erwähnt. Bei UV-Bestrahlung von $C_6H_6Cr(CO)_3$ in Methanol bildet sich $Cr(OCH_3)_3$ [102], während bei der Umsetzung von $C_6H_5MeCr(CO)_3$ mit $NO[PF_6]$ in Acetonitril nicht – wie erwartet – das Kation $[C_6H_5MeCr(CO)_2NO]^+$ sondern $[Cr(NO)_2(CH_3CN)_4]^{2+}$ entsteht [103].

Der Ringaustausch Aromat gegen Aromat gemäß (34)

$$ArM(CO)_3 + Ar' \longrightarrow Ar'M(CO)_3 + Ar \qquad (34)$$

wurde von G. Natta und Mitarbeitern [104] bereits 1958, d.h. kurz nach der erstmaligen Darstellung der Aromaten-metall-tricarbonyle, für M = Cr untersucht und später zur Synthese von Komplexen $Ar'Cr(CO)_3$ mit $Ar' = C_6H_5X$ und $p\text{-}MeC_6H_4X$ (X = Me, OMe, SMe, NMe_2, COMe, COOMe) und $Ar' = C_4H_4S$ sowie verschiedenen substituierten Thiophenderivaten genutzt [105,106]. Als Ausgangsverbindung $ArCr(CO)_3$ diente dabei Benzol- oder Toluol-chrom-tricarbonyl. Bei der notwendigen hohen Reaktionstemperatur von 180–220 °C ist ein Arbeiten im geschlossenen System und die Verwendung eines möglichst großen Überschusses von Ar′ günstig. Komplexe mit ^{14}C-markierten Aromaten wurden für M = Cr, Mo, W und Ar′ = ^{14}C-Benzol, ^{14}C-Toluol und ^{14}C-Naphthalin ebenfalls gemäß (34) synthetisiert; dabei hat sich sowohl die Zufuhr thermischer als auch photochemischer Energie bewährt [107,108]. Bei Bestrahlung mit Licht der Wellenlänge $\lambda = 3660$ Å erfolgt der Ringaustausch für $C_6H_5MeW(CO)_3$ sehr viel langsamer als für die entsprechenden Chrom- und Molybdän-Komplexe [108]. Vorstellungen über den Mechanismus dieses Ringaustausches werden unter 3. diskutiert.

Über Umsetzungen kationischer Aromaten-Metalltricarbonyl-Komplexe mit Lewis-Basen ist wenig bekannt. $[s\text{-}C_6H_3Me_3Mn(CO)_3]^+$ reagiert mit Diäthylentriamin (dien) bei 60 °C ohne Solvens [100] und mit Tris-1.1.1-(dimethylarsinomethyl)äthan (triars) in siedendem Cyclohexan [109] zu $[(dien)Mn(CO)_3]^+$ bzw. $[(triars)Mn(CO)_3]^+$.

Eine Spaltung der Ring-Metall-Bindung wird ebenfalls bei den Reaktionen der aus $C_6Me_6M(CO)_3$ (M = Mo, W) und $SbCl_5$ dargestellten Kationen $[C_6Me_6M(CO)_3Cl]^+$ mit Aminen, Phosphinen und Arsinen beobachtet. Mit einzähnigen Liganden L wie z.B. PPh_3 oder $AsPh_3$ entstehen unter Disproportionierung – neben $C_6Me_6M(CO)_3$ – die ungeladenen Verbindungen $M(CO)_3L_2Cl_2$; mit zweizähnigen Liganden L′ wie z.B. dipy, phen oder o-$C_6H_4(AsMe_2)_2$ werden dagegen unter Freisetzung von C_6Me_6 und einer CO-Gruppe die Kationen $[M(CO)_2L'_2Cl]^+$ erhalten [110]. Komplexe wie z.B. $W(CO)_3(PPh_3)_2Cl_2$ haben Bedeutung als CO-Überträger erlangt [111]. Bei der Umsetzung von $[C_6Me_6M(CO)_3Cl]^+$ mit dreizähnigem Bis-(o-dimethylarsinophenyl)-methylarsin (o-triars) findet ein „normaler" Austausch unter Bildung von $[M(CO)_3(\text{o-triars})Cl]^+$ statt [110].

Auf folgende Analogie sei noch hingewiesen: Die Hexaalkylborazol-chromtricarbonyle $R_3B_3N_3R'_3Cr(CO)_3$ gehen ebenso wie die isoelektronischen Benzol-Komplexe $C_6R_6Cr(CO)_3$ sehr leicht Ringligandenverdrängungsreaktionen ein. Mit PH_3 [112], tertiären Phosphiten $P(OR)_3$ (R = Me, Et, But, Ph), Trialkylphosphinen PR_3 (R = Et, But), Phenyldiäthylphosphin $PPhEt_2$ und Isonitrilen CNR (R = C_6H_{11}, Ph) entstehen bereits bei 0 °C gemäß (35) die cis-Tricarbonyl-Verbindungen. Mit PPh_3 und $P(C_6H_{11})_3$ werden – vermutlich aus sterischen Gründen – die trans-Tetracarbonyle $Cr(CO)_4(PR_3)_2$ gebildet [113,114].

$$R_3B_3N_3R'_3Cr(CO)_3 + 3\,L \longrightarrow \text{cis-}Cr(CO)_3L_3 + R_3B_3N_3R'_3 \qquad (35)$$

π-Donoren wie z.B. Benzol, Mesitylen oder Cycloheptatrien führen bei Umsetzung mit $B_3N_3Me_6Cr(CO)_3$ bei 50–80 °C ebenfalls zu einem Ringaustausch [115]. Die Bereitwilligkeit, mit der diese Reaktionen verlaufen, läßt die Möglichkeit offen, daß ausgehend von den Borazol-Halbsandwich-Komplexen unter geeigneten Bedingungen auch die Darstellung sonst nicht zugänglicher oder thermodynamisch wenig stabiler, substituierter Metallcarbonyle gelingt.

3. Kinetische Untersuchungen

Nach dem unter 2. gegebenen Überblick über die präparativen Ergebnisse erhebt sich die Frage, ob es für die Vielfalt der beschriebenen Reaktionen aus mechanistischer Sicht einen gemeinsamen Aspekt gibt. Mit aller gebotenen Vorsicht – die Zahl der kinetischen Untersuchungen über Ligandenverdrängungsreaktionen ist noch relativ gering – darf diese Frage mit einem „Ja" beantwortet werden.

Zunächst eine Zusammenstellung der bisher vorliegenden Resultate:

Dicyclopentadienylnickel reagiert (wie unter 2.1. bereits erwähnt) mit Trialkylphosphiten $P(OR)_3$ zu $Ni[P(OR)_3]_4$. Die Abnahme der Konzentration von $Ni(C_5H_5)_2$ (gemessen am Absorptionsmaximum von $Ni(C_5H_5)_2$ bei 690 mm) folgt für R = Me, Et und But gemäß (36) einem Geschwindigkeitsgesetz 3. Ordnung, d.h. 1. Ordnung in bezug auf $[Ni(C_5H_5)_2]$ und 2. Ordnung in bezug auf $[P(OR)_3]$ [6,116].

$$-\frac{d[Ni(C_5H_5)_2]}{dt} = k[Ni(C_5H_5)_2][P(OR)_3]^2 \qquad (36)$$

Die Geschwindigkeitskonstanten k und auch die Aktivierungsparameter sind nahezu unabhängig von der Art des Alkylrestes; sie unterscheiden sich ebenso bei Änderung des Lösungsmittels praktisch nicht (Tabelle 3).

Tabelle 3. *Kinetische Daten der Reaktion von $Ni(C_5H_5)_2$ und $P(OR)_3$*

Phosphit	Solvens	k (60 °C) [$sec^{-1} M^{-2}$]	E_a [Kcal/Mol]	$\Delta S^{\neq}$ [e.u.]
$P(OMe)_3$	Dioxan	$10.5 \cdot 10^{-4}$	7.5	-51.8
$P(OEt)_3$	Dioxan	$11.6 \cdot 10^{-4}$	6.8	-53.8
$P(OBut)_3$	Dioxan	$10.6 \cdot 10^{-4}$	6.7	-54.0
$P(OEt)_3$	Methylcyclohexan	$11.0 \cdot 10^{-4}$		
$P(OEt)_3$	Toluol	$8.0 \cdot 10^{-4}$		
$P(OEt)_3$	Diisopropyläther	$11.5 \cdot 10^{-4}$		

Die mechanistische Interpretation der kinetischen Daten der Reaktion von $Ni(C_5H_5)_2$ und $P(OR)_3$ geht davon aus [6], daß sich in Einklang mit den Vorstellungen von Ustynyuk und Mitarbeitern [14,15] zunächst ein 1 : 1 Komplex (A) bildet, der dann mit überschüssigem Phosphit weiterreagiert. Für die – sicher wenig stabile – Zwischenverbindung (B) kommt eine Zusammensetzung entsprechend $(C_5H_5)_2Ni[P(OR)_3]_2$ oder $C_5H_5Ni[P(OR)_3]_2$ in Betracht.

$$Ni(C_5H_5)_2 + P(OR)_3 \underset{k_{-1}}{\overset{k_1}{\rightleftharpoons}} \underset{(A)}{[(C_5H_5)_2NiP(OR)_3]} \xrightarrow[+P(OR)_3]{k_2}$$

$$(B) \xrightarrow[+P(OR)_3]{k_3} (C) \longrightarrow \cdots \xrightarrow[+P(OR)_3]{} Ni[P(OR)_3]_4 \qquad (37)$$

Bei Annahme einer Reaktionsfolge nach (37) führt die Anwendung des Bodenstein-Theorems ("steady-state approximation") zu Gl. (38), die für $k_{-1} \gg k_2[P(OR)_3]$ mit den experimentellen Werten übereinstimmt.

$$-\frac{d[\mathrm{Ni(C_5H_5)_2}]}{dt} = \frac{k_1 k_2[\mathrm{Ni(C_5H_5)_2}][\mathrm{P(OR)_3}]^2}{k_{-1} + k_2[\mathrm{P(OR)_3}]} \tag{38}$$

Die Existenz der 1 : 1-Zwischenverbindung (A) ist durch spektroskopische Messungen belegt [12]; ob sie allerdings die von Ustynyuk [14,15] vorgeschlagene Struktur gemäß $(\pi\text{-}C_5H_5)Ni(\sigma\text{-}C_5H_5)P(OR)_3$ besitzt, ist noch nicht gesichert.

Die Umsetzung des Quasi-Sandwich-Komplexes $(\pi\text{-}C_5H_5)Ni(\pi\text{-}C_3H_5)$ mit Trialkylphosphiten zu $Ni[P(OR)_3]_4$ erfolgt ebenfalls – analog zu (36) – nach einem Zeitgesetz 3. Ordnung. Für die entsprechende Reaktion von $(\pi\text{-}C_5H_5)Pd(\pi\text{-}C_3H_5)$ mit $P(OPh)_3$ wird dagegen eine Beziehung gemäß (39) gefunden [116].

$$-\frac{d[\mathrm{C_5H_5PdC_3H_5}]}{dt} = k\,[\mathrm{C_5H_5PdC_3H_5}][\mathrm{P(OR)_3}] \tag{39}$$

Der Hauptgrund für den Unterschied in der Ordnung der Reaktionen der homologen π-Allyl-π-Cyclopentadienyl-Nickel- und Palladium-Komplexe mit $P(OR)_3$ dürfte in der unterschiedlichen Stabilität der primär gebildeten Zwischenverbindung zu suchen sein, die für M = Pd offensichtlich sehr rasch mit $P(OR)_3$ weiterreagiert. Bei einem zu (37) analogen Reaktionsverlauf würde dann gelten $k_2[P(OR)_3] \gg k_{-1}$, d.h. die Konstante k von Gl. (39) würde der Geschwindigkeitskonstante k_1 entsprechen.

Über die Kinetik des Ringligandenaustausches in *Aromaten-metall-tricarbonylen* gemäß (34) liegen ausführliche Untersuchungen von W. Strohmeier und Mitarbeitern [117–119] vor. Für Ar = Benzol, Toluol und Chlorbenzol und Ar' = $^{14}C\text{-}C_6H_6$, $^{14}C\text{-}C_6H_5Me$ und $^{14}C\text{-}C_6H_5Cl$ resultiert ein additives Geschwindigkeitsgesetz:

$$\frac{d[\mathrm{Ar'M(CO)_3}]}{dt} = k[\mathrm{ArM(CO)_3}]^2 + k'[\mathrm{ArM(CO)_3}][\mathrm{Ar'}] \tag{40}$$

Die Konstanten k und k', die eine Funktion der Hammet'schen σ-Parameter sind, nehmen in der Reihe Cr < W < Mo zu. Ist Ar und Ar' ein kondensierter Aromat, wie z.B. Naphthalin (oder auch ein Triolefin wie z.B. Cycloheptatrien), so gehorcht die Bildung der markierten Verbindungen $Ar'M(CO)_3$ der Gl. (41):

$$\frac{d[\mathrm{Ar'M(CO)_3}]}{dt} = k[\mathrm{ArM(CO)_3}] + k'[\mathrm{ArM(CO)_3}][\mathrm{Ar'}] \tag{41}$$

Der Austausch scheint also stets nach zwei verschiedenen, parallel verlaufenden Mechanismen zu erfolgen. Dabei ist vor allem bemerkenswert, daß bei den Reaktionen der Benzol-metall-tricarbonyle die Bildung eines aktivierten Komplexes XII aus 2 Molekülen $ArM(CO)_3$ (X = H, Me, Cl) anzunehmen ist.

2 $ArM(CO)_3$ 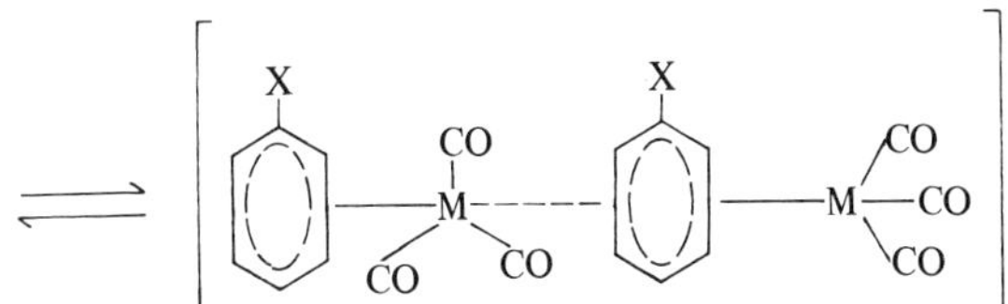

XII

$$\longrightarrow Ar + ArM(CO)_3 \ [\ + M(CO)_3 \xrightarrow{+Ar'} Ar'M(CO)_3\ ] \qquad (42)$$

Verbindungen $ArM(CO)_3$, in denen der Aromat Ar in der cis- oder trans-Konfiguration gebunden sein kann, sollten dann bei Energiezufuhr auch in Abwesenheit eines freien Liganden Ar oder Ar′ isomerisieren, was jedoch im Fall 1- und 2-substituierter Indane nicht beachtet wurde [120].

Die Umsetzungen der *Aromaten-molybdän- und -wolfram-tricarbonyle* mit Phosphinen und Phosphiten nach (35) folgen einem „einfachen" Geschwindigkeitsgesetz 2. Ordnung:

$$\frac{d[M(CO)_3L_3]}{dt} = k[ArM(CO)_3]\,[L] \qquad (43)$$

Wie für den Ringligandenaustausch ist die Geschwindigkeit bei einer bestimmten Temperatur für M = Mo größer als für M = W; bei Änderung von Ar nehmen die k-Werte in der Reihenfolge $C_6H_6 > C_6H_5Me > p\text{-}C_6H_4Me_2 > s\text{-}C_6H_3Me_3$ ab. Sowohl Basolo [93] als auch Pidcock [95,96] postulieren für die Reaktion von $ArM(CO)_3$ und L (siehe (44)) als primären, geschwindigkeitsbestimmenden Schritt einen Angriff von L auf das Metallatom und eine nachfolgende sukzessive Spaltung der Aromat-Metall-Bindung. Für die zunächst gebildete Zwischenverbindung XIII steht eine dien-artige Fixierung des Aromaten (als quasi-zweizähniger Ligand) zur Diskussion. Die Annahme k_2 und $k_3 \gg k_1$ gründet sich auf die Tatsache, daß die Bindung zwischen einem Di- oder Oligoolefin und einem Metall allgemein sehr viel leichter gespalten wird als eine Bindung Aromat-Metall. Die Cycloheptatrien-metall-tricarbonyle reagieren z.B. mit Trimethylphosphit um mehrere Größenordnungen schneller als die entsprechenden Benzol-Komplexe; die Aktivierungsenergie beträgt für die Umsetzung von $C_7H_8Mo(CO)_3$ und $P(OMe)_3$ (in Methylcyclohexan) 9.8 Kcal/Mol [121], für die Reaktion von $s\text{-}C_6H_3Me_3Mo(CO)_3$ und $P(OMe)_3$ (in 1.2-Dichloräthan) dagegen 17.3 Kcal/Mol [95].

Besonderes Interesse verdient der Befund, daß Verbindungen mit Element-Sauerstoff-Doppelbindungen die Ligandenverdrängung beeinflussen können. Katalytische Mengen von Trimethylphosphat $(MeO)_3PO$, Dimethylformamid Me_2NCHO oder Dimethylsulfoxid Me_2SO bewirken z.B. eine deutliche Erhöhung der Geschwindigkeit der Umsetzung von $ArM(CO)_3$ und $P(OMe)_3$ [122].

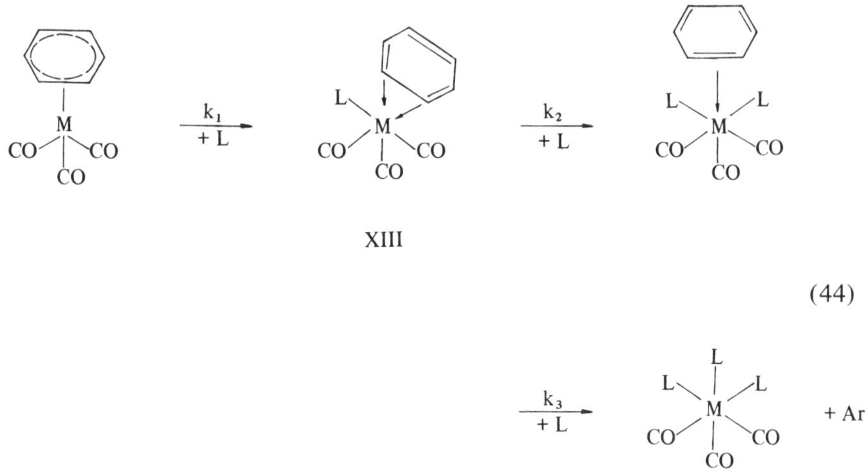

XIII

(44)

Die Bildung von cis-$M(CO)_3[P(OMe)_3]_3$ (als dem nach wie vor einzigsten Reaktionsprodukt) gehorcht in diesem Fall der Gl. (45) [K = Katalysator]:

$$\frac{d[M(CO)_3(P(OMe)_3)_3]}{dt} = k[ArM(CO)_3]\,[P(OMe)_3] + k_K\,[ArM(CO)_3]\,[K] \tag{45}$$

Das Ausmaß der Katalyse ist für M = W gravierender als für M = Mo. Für die Reaktion von $C_6H_5COOMeW(CO)_3$ und $P(OMe)_3$ resultieren bei 50 °C in 1.2-Dichloräthan Geschwindigkeitskonstanten k_K, die um einen Faktor 30 bis 300 größer sind als die Konstante k.

Im Hinblick auf den Mechanismus der Ligandenverdrängung ist der 2. Term von Gl. (45) so interpretiert worden, daß im *Primärschritt* eine Wechselwirkung von $ArM(CO)_3$ und K erfolgt und danach, und zwar *vor* einer vollständigen Spaltung der Aromat-Metall-Bindung, sich eine Substitution von K durch $P(OMe)_3$ anschließt [122]. Daß die katalytisch wirkenden X=O-Verbindungen leichter zu einem nucleophilen Angriff auf den Komplex $ArM(CO)_3$ als das Trimethylphosphit fähig sind (siehe die Größe von k und k_K), könnte auf sterische Ursachen zurückzuführen sein: Die Knüpfung einer Bindung M–O–P$\lessgtr$ ist weniger gehindert als die Knüpfung einer Bindung M–P$\lessgtr$.

Die Reaktionen der *Hexaalkylborazol-chrom-tricarbonyle* $R_3B_3N_3R'_3Cr(CO)_3$ mit tertiären Phosphiten $P(OR)_3$ verlaufen nach einem Geschwindigkeitsgesetz 2. Ordnung [114,123]. Nimmt man einmal an, daß das für die Umsetzungen der Benzol-Molybdän- und Benzol-Wolfram-Komplexe $C_6R_6M(CO)_3$ mit L = PR_3 und $P(OR)_3$ gefundene Geschwindigkeitsgesetz auch für die Reaktionen der entsprechenden Chrom-Komplexe gilt, so ist die durch die Isosterie der Verbindungen $R_3B_3N_3R'_3Cr(CO)_3$ – $C_6R_6Cr(CO)_3$ gegebene Analogiebeziehung auch aus kinetischer Sicht gewahrt. Sehr große Unterschiede bestehen allerdings in der

Größe der Geschwindigkeitskonstanten der Reaktionen der Benzol- und Borazol-Chrom-Komplexe mit L. Während z.B. $C_6H_6Cr(CO)_3$ mit $P(OPh)_3$ erst oberhalb 200 °C mit gut meßbarer Geschwindigkeit reagiert [97] (die entsprechende Umsetzung von $C_6Me_6Cr(CO)_3$ wäre dann wahrscheinlich, nach den Untersuchungen am Molybdän zu schließen, als noch langsamer zu erwarten), ist die Reaktion von $B_3N_3Me_6Cr(CO)_3$ mit $P(OEt)_3$ in Cyclohexan oder 1.2-Dichloräthan bei 30 °C und einem Molverhältnis 1 : 10 bereits nach 15–20 Minuten beendet. Die Aktivierungsenergie für die Spaltung der Borazol-Metall-Bindung ist außerordentlich gering und beträgt für die Umsetzung von $B_3N_3Me_6Cr(CO)_3$ und $P(OEt)_3$ in Cyclohexan 5.95, in 1.2-Dichloräthan 8.9 Kcal/Mol [114].

Ähnlich wie im Fall der Aromaten-Komplexe $C_6R_6M(CO)_3$ (M = Mo, W) hängt auch bei den Hexaalkylborazol-chrom-tricarbonylen die Geschwindigkeit der Ligandenverdrängung deutlich von den Substituenten R und R′ ab. $B_3N_3Et_6Cr(CO)_3$ reagiert z.B. mit $P(OEt)_3$ bei 20 °C in Cyclohexan 40mal langsamer als $B_3N_3Me_6Cr(CO)_3$; die Aktivierungsenergie ist für die Umsetzung der Hexaäthyl-Verbindung um mehr als 4 Kcal/Mol größer und beträgt 10.2 Kcal/Mol [123].

Faßt man die Ergebnisse der kinetischen Untersuchungen zusammen, so ist in bezug auf ihre mechanistische Interpretation als wichtigster gemeinsamer Aspekt festzuhalten, daß der Primärschritt der Ligandenverdrängung offensichtlich stets in einem *Angriff der substituierenden Lewis-Base L auf den Komplex* besteht. Diese Tatsache, daß der Ligandenaustausch durch einen bimolekularen, assoziativen Vorgang eingeleitet wird, verdient insofern besondere Beachtung, als Sandwich- oder Halbsandwich-Verbindungen wie $M(C_5H_5)_2$ oder $C_6H_6M(CO)_3$ als quasi-oktaedrisch koordiniert angesehen werden können und es durch eine große Zahl von Arbeiten belegt ist, daß oktaedrische Komplexe „klassischer Prägung" (wie z.B. $[Co(NH_3)_5X]^{2+}$ oder $[Cr(en)_2X_2]^+$), aber auch oktaedrisch gebaute Metallcarbonyle (wie z.B. $Cr(CO)_6$ oder $Mn(CO)_5X$ [124]) vorwiegend nach einem dissoziativen (S_N1-) Mechanismus reagieren.

Die Frage, ob die Ursache für den bevorzugten assoziativen Primärschritt bei den hier diskutierten Ligandenverdrängungsreaktionen in erster Linie in elektronischen oder aber in sterischen Einflüssen zu suchen ist, läßt sich nicht eindeutig beantworten. Auf folgende Punkte sei hingewiesen:

1) *Aromatische Ringsysteme* wie C_5H_5 oder C_6H_6 können als formal dreizähnige Liganden gelten; sie sind im allgemeinen fester gebunden als typisch einzähnige Liganden wie CO oder PR_3. Eine Dissoziation, d.h. eine monomolekulare Spaltung einer C_5H_5-M- oder C_6H_6-M-Bindung, sollte daher einen relativ hohen Energiebetrag erfordern und aus diesem Grund wenig begünstigt sein – auch wenn man bedenkt, daß der Bruch einer Ring-Metall-Bindung möglicherweise sukzessive erfolgt.Der Verlust an Resonanzenergie, z.B. beim Übergang von einem aromatisch gebundenen C_6H_6 zu einem dien-artig gebundenen C_6H_6, wäre in diesem Fall für die Energiebilanz wahrscheinlich bestimmend. Die Annahme ist somit berechtigt, daß eine Anlagerung von L an den Kom-

plex aus energetischer Sicht eher zu realisieren ist als die Spaltung einer C_nH_n-Metall-Bindung, d.h. daß bei den Ligandenverdrängungsreaktionen ein assoziativer Schritt gegenüber einem dissoziativen Schritt dominiert.

2) Bei einer Betrachtung der *sterischen Einflüsse* könnte man davon ausgehen, daß ein Angriff von L auf einen Komplex wie $Ni(C_5H_5)_2$ oder $C_6H_6Mo(CO)_3$ in einer zu den Ringebenen parallelen Ebene erfolgt (siehe XIV und XV). Es erscheint plausibel, daß hier – im Gegensatz zu streng oktaedrischen Verbindungen wie z.B. $Cr(CO)_6$ – eine Wechselwirkung in der angegebenen Richtung bevorzugt ist.

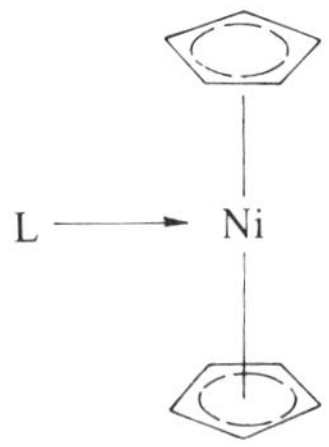

XIV

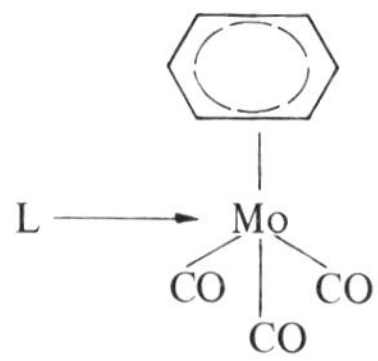

XV

Fast alle Autoren der in diesem Abschnitt angeführten Arbeiten haben daher auch bei der Diskussion ihrer Vorschläge zum Verlauf von Ligandenverdrängungsreaktionen diese Möglichkeit in den Vordergrund gerückt. Ist diese Deutung des Primärschritts jedoch wirklich die einzigste und ist sie die beste?

Betrachtet man einmal maßstabgerechte Molekülmodelle von Verbindungen wie z.B. $M(CO)_6$ und $ArM(CO)_3$, so fällt es schwer einzusehen, daß das Metall in dem Halbsandwich-Komplex *weniger abgeschirmt,* d.h. einem Angriff von L *leichter zugänglich* sein soll als in dem Metallhexacarbonyl. Eine Alternative zu der in XIV und XV skizzierten Vorstellung wäre die, daß sich der primäre Angriff von L nicht auf das Metall *sondern auf den abzuspaltenden Ringligand richtet* (siehe XVI und XVII).

XVI

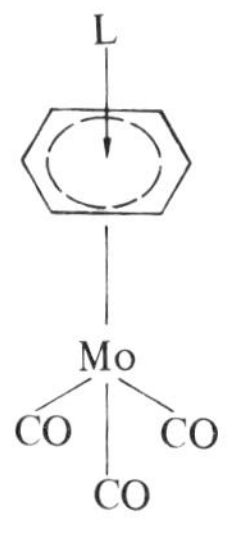

XVII

Als Stütze für diesen Vorschlag (der von uns erstmals im Zusammenhang mit der Diskussion des Mechanismus der Reaktion von $B_3N_3Me_6Cr(CO)_3$ und

$P(OR)_3$ geäußert und der später auch von White und Mawby [71] für die Umsetzung von $C_5H_5Mo(CO)_3Cl$ und PR_3 in Betracht gezogen wurde) könnten folgende Argumente dienen:

a) Nach Aussage von IR-Messungen besitzt der Ringligand in den Komplexen $C_6R_6M(CO)_3$ [125] und $R_3B_3N_3R'_3M(CO)_3$ [113,126] einen stärkeren Donor- als Akzeptorcharakter, d.h. es liegt eine Ladungsverteilung gemäß Ring $^{\delta+}$-$M(CO)_3^{\delta-}$ vor. Dipolmessungen [127–129] wie auch die Art der Substituenteneinflüsse auf die Geschwindigkeit des Austausches von $C_6H_5XM(CO)_3$ und ^{14}C-C_6H_5X [119] stehen damit in Einklang. Es wäre denkbar, daß eine positive Partialladung am Ring die Wechselwirkung mit einem nucleophilen Reaktionspartner begünstigt und durch diese Wechselwirkung eine Schwächung der Ring-Metall-Bindung resultiert.

b) Die für verschiedene Ligandenverdrängungsreaktionen ermittelten Aktivierungsparameter erscheinen in zweifacher Hinsicht bermerkenswert: Die Werte der Aktivierungsenergie sind relativ klein; die Werte der Aktivierungsentropie sind stark negativ. Ersteres könnte als ein Hinweis auf einen *wenig gehinderten* Angriff von L auf den Komplex, letzteres als Zeichen für einen *hoch geordneten* Übergangszustand gelten. Eine ähnliche Größenordnung der E_a- und $\Delta S^{\neq}$-Werte findet man auch bei einigen S_N2-Substitutionsreaktionen quadratisch-planarer Komplexe, z.B. von Pt(II) und Au(III), in denen das Metall oberhalb und unterhalb der Ligandenebene wenig abgeschirmt und daher einem Angriff von L aus der Richtung senkrecht dazu leicht zugänglich ist.

c) Zahlreiche physikalische Messungen haben gezeigt, daß in Sandwich-Komplexen wie $M(C_5H_5)_2$ oder $M(C_6H_6)_2$ das Ausmaß der Metall-C_nH_n-Rückbindung stärker und damit die Elektronendichte an den Ringligandenatomen größer als im Fall der Cyclopentadienyl- und Aromaten-metall-carbonyle ist. Danach wäre zu erwarten, daß bei einer primären Wechselwirkung zwischen L und C_nH_n die Geschwindigkeit der Ligandenverdrängung nicht unbedingt proportional zu der Donorstärke von L zunimmt, sondern daß möglicherweise solche Lewis-Basen L, die gute Akzeptoreigenschaften besitzen, bevorzugt mit den Sandwich-Komplexen $M(C_nH_n)_2$ reagieren. Einige qualitative Befunde könnten diese Vermutung bestätigen: Die Umsetzung von $Ni(C_5H_5)_2$ mit PPh_3 verläuft in Dioxan bei 50 °C etwa um einen Faktor 10^2 langsamer als mit $P(OEt)_3$ [12]. Bei Verwendung eines größeren Ligandenüberschusses genügt für die Bildung von $Ni(CNPh)_4$ aus $Ni(C_5H_5)_2$ und CNPh eine Reaktionstemperatur von 0 °C, dagegen ist für diejenige von $Ni(PPh_3)_4$ aus $Ni(C_5H_5)_2$ und PPh_3 eine solche von 60–120 °C notwendig [4]. $Cr(C_6H_6)_2$ reagiert unter energischen Bedingungen zwar mit CO und PF_3 (die beide sehr gute Akzeptoren sind), jedoch nicht mit PPh_3, bei dem die Donoreigenschaften überwiegen.

Es sei ausdrücklich betont, daß der Vorschlag einer Wechselwirkung zwischen dem substituierenden Liganden L und dem Ring C_nH_n im Primärschritt der hier diskutierten Ligandenverdrängungsreaktionen vorerst nicht mehr als *eine Möglichkeit* darstellt, deren Relevanz weitere Untersuchungen klären müssen. Die Alternative – Angriff von L auf das Metall oder auf den Ringligand – sollte insgesamt, unter Berücksichtigung der die Energie des Übergangszustandes bestimmenden Reaktionsparameter, eher im Sinne eines „sowohl – als auch" als eines „entweder – oder" gesehen werden. Von Interesse dürfte in diesem Zusammenhang noch sein, daß Sandwich-Komplexe wie $M(C_5H_5)_2$ (M = Fe, Ru, Os, Co, Ni) [130] und $Cr(C_6H_6)_2$ [131], aber auch Halbsandwich-Komplexe wie $ArM(CO)_3$ [132], mit organischen π-Akzeptorverbindungen wie z.B. *s*-Trinitrobenzol 1 : 1-oder 1 : 2-Charge-Transfer-Addukte bilden, in denen der C_nH_n-Ringligand als π-Donor fungiert und eine direkte Wechselbeziehung zwischen diesem und dem π-Akzeptor besteht.

In Anbetracht der zuletzt und auch oben unter c) angeführten Resultate erscheint schließlich die Frage berechtigt, ob es überhaupt zutreffend ist (wie es in den meisten Publikationen getan wird), den Primärschritt bei der Umsetzung von L mit einen C_nH_n-Metall-Komplex als einen *nucleophilen* Angriff zu interpretieren. Wäre es nicht vielleicht sinnvoller, z.B. gerade bei den Reaktionen der Sandwich-Verbindungen mit Akzeptorliganden wie CO, PF_3 oder $P(OR)_3$ von einem *elektrophilen* Angriff zu sprechen? Die Hauptschwierigkeit, hier zu einer klaren Unterscheidung zu kommen, besteht sicher darin, die Donor- und Akzeptoreigenschaften eines Liganden L eindeutig voneinander zu trennen. Es wäre durchaus denkbar, daß in Abhängigkeit von der elektronischen Struktur eines C_nH_n-Metall-Komplexes der gleiche Ligand L in einem Fall als vorwiegend nucleophiler Reaktionspartner, im anderen Fall dagegen als vorwiegend elektrophiler Reaktionspartner wirkt. Dies würde heißen, daß einmal die Donoreigenschaften, zum anderen die Akzeptoreigenschaften von L für die Reaktion maßgebend sind. Auf diese Möglichkeit der *„biphilic nature"* von L und auch auf die Schwierigkeit, Substitutionsreaktionen von Metallkomplexen eindeutig nach „S_N" oder „S_E" zu klassifizieren, haben bereits Langford und Gray in ihrer Monographie „Ligand Substitution Processes" kurz aufmerksam gemacht [133]. Im Rahmen der Diskussion der Mechanismen von Ringligandenverdrängungsreaktionen von Sandwich- und Halbsandwich-Verbindungen sei auf diese Problematik noch einmal hingewiesen.

4. Liganden-Übertragungsreaktionen

Die Spaltung einer C_nH_n-M-Bindung eines Aromaten-Metall-Komplexes kann nicht nur durch Reaktion mit einer *freien, nicht-koordinierten* Lewis-Base (im vorhergehenden allgemein mit L bezeichnet) erfolgen, sondern ebenso durch Wechselwirkung mit einem *zweiten Metallkomplex*. Dabei findet – im Idealfall – die Übertragung eines Liganden A (der Zusammensetzung C_nH_n) von dem Komplex MA_m auf das Metall M′ der Verbindung $M'B_n$ und gleichzeitig die Übertragung eines – oder mehrerer – Liganden B (nicht unbedingt, aber möglicherweise auch der Zusammensetzung C_nH_n) von dem Komplex $M'B_n$ auf das Metall M der Verbindung MA_m statt.

$$MA_m + M'B_n \longrightarrow MA_{m-x}B_y + M'B_{n-y}A_x \qquad (46)$$

Die Frage, ob und in welchem Ausmaß dem eigentlichen Ligandentransfer eine Dissoziation einer M-A- oder M'-B-Bindung vorausgeht und dann eine freie Lewis-Base A oder B mit dem Komplex MA_m bzw. $M'B_n$ reagiert, kann nicht immer eindeutig beantwortet werden. Kinetische Untersuchungen, die zu einer Lösung dieses Problems beitragen könnten, stehen noch aus.

Umsetzungen des in (46) angegebenen Typs (die teilweise auch als Komplex-Disproportionierungs- oder -Synproportionierungs-Reaktionen, im Englischen als „redistribution-", „ligand-migration-" oder „scrambling-reactions" bezeichnet werden) sind teilweise schon kurz unter 2. erwähnt worden. Eines der ersten in der Literatur beschriebenen Beispiele für eine solche Reaktion ist die Darstellung von Benzol-chrom-tricarbonyl gemäß (47) [81].

$$Cr(C_6H_6)_2 + Cr(CO)_6 \xrightarrow{C_6H_6} 2\ C_6H_6Cr(CO)_3 \qquad (47)$$

Die angewendeten Reaktionsbedingungen (Arbeiten im Einschlußrohr bei 220 °C) und die nur mäßige Ausbeute (~ 25 %) lassen es allerdings nicht als sicher erscheinen, ob hierbei wirklich eine Ligandenübertragung und nicht eine Reaktion von $Cr(CO)_6$ mit dem Lösungsmittel Benzol erfolgt.

Bedenken dieser Art treffen sicher nicht zu bei der Bildung der Komplexe $[C_5H_5NiCO]_2$ [25], $C_5H_5Ni(CO)_2Fe(CO)C_5H_5$ [134], $[C_5H_5NiCNPh]_2$ [3], $C_5H_5Ni(PR_3)X$ [135-137] und $C_5H_5TiCl_3$ [138] gemäß (48)–(52).

$$Ni(C_5H_5)_2 + Ni(CO)_4 \xrightarrow[80\ °C]{C_6H_6} C_5H_5Ni\genfrac{}{}{0pt}{}{CO}{CO}NiC_5H_5 + 2\ CO \qquad (48)$$

$$Ni(C_5H_5)_2 + Fe(CO)_5 \xrightarrow[80\ °C]{C_6H_6} C_5H_5Ni\genfrac{}{}{0pt}{}{CO}{CO}Fe(CO)C_5H_5 + 2\ CO \qquad (49)$$

$$Ni(C_5H_5)_2 + Ni(CNPh)_4 \xrightarrow[80\ °C]{C_6H_6} C_5H_5(PhNC)Ni\text{-}Ni(CNPh)C_5H_5 + 2\ CNPh \qquad (50)$$

$$Ni(C_5H_5)_2 + (PR_3)_2NiX_2 \xrightarrow[65\ °C]{THF} 2\ C_5H_5Ni(PR_3)X \qquad (51)$$

$$(C_5H_5)_2TiCl_2 + TiCl_4 \xrightarrow[140\ °C]{Xylol} 2\ C_5H_5TiCl_3 \qquad (52)$$

Die Umsetzung von $Ni(C_5H_5)_2$ und $Ni(CO)_4$ führt vor allem bei höherer Temperatur (~ 120 °C), und zwar auch im Sinne einer Ligandenübertragung, neben $[C_5H_5NiCO]_2$ zu dem sehr stabilen, dreikernigen Komplex $(C_5H_5)_3Ni_3(CO)_2$ [25]; weiterhin bilden sich bei der Reaktion von $Ni(C_5H_5)_2$ und $Fe(CO)_5$ neben $C_5H_5Ni(CO)_2Fe(CO)C_5H_5$ noch $[C_5H_5Fe(CO)_2]_2$ und geringe Mengen $[C_5H_5NiCO]_2$ [134].

Ligandenübertragungen zwischen jeweils zwei Fünfring-Komplexen und zwei Sechsring-Komplexen sind ebenfalls bekannt. Alkylierte Ferrocene (aber auch Verbindungen $C_5H_5FeC_5H_4X$ mit X = OR und SR [38,139])) synproportionieren z.B. in Gegenwart von $AlCl_3$ gemäß:

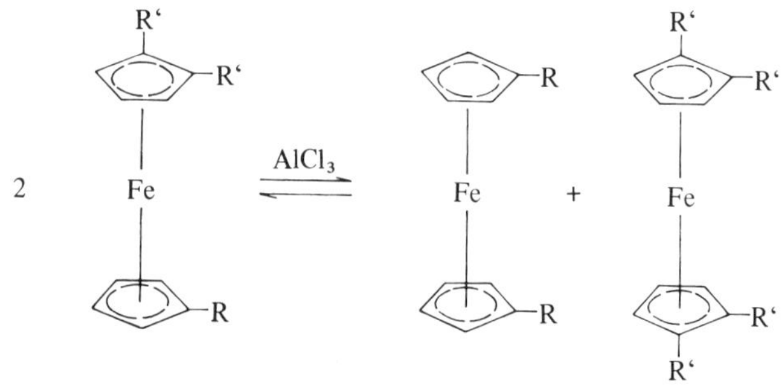

(53)

[R = H, Et; R' = Et] XVIII XIX

Die Tatsache, daß bei der nach (53) formulierten Reaktion praktisch ausschließlich die Verbindungen XVIII und XIX – und zwar zu gleichen Teilen – entstehen, weist darauf hin, daß für ihre Bildung nicht eine Wanderung der Substituenten R und R′ sondern eine Spaltung der Metall-Ring-Bindungen maßgebend ist [42]. Mit zunehmender Zahl von R und R′ wird die Ligandenübertragung erheblich erschwert; daher ist z.B. Nonaäthylferrocen immun gegenüber einer Synproportionierung in Octa- und Deca-äthylferrocen [42].

Daß die Ligandenübertragungsreaktionen wahrscheinlich immer *Gleichgewichtsreaktionen* sind, haben nicht nur die Untersuchungen von Bublitz [42] sondern auch diejenigen von Hein [89] an Diaromaten-Chrom-Komplexen gezeigt. Dibenzolchrom und Bis(diphenyl)chrom sind ausgehend von Benzol-diphenylchrom zugänglich.

$$2\ Cr(C_6H_6)(C_6H_5C_6H_5) \xrightleftharpoons{AlCl_3} Cr(C_6H_6)_2 + Cr(C_6H_5C_6H_5)_2 \qquad (54)$$

Eine *vollständige* Umwandlung in die reinen Diaromaten-Komplexe konnte auch durch 20-stündiges Erhitzen im Einschlußrohr nicht erreicht werden. Die Bildung des gemischten Komplexes $Cr(C_6H_6)(C_6H_5C_6H_5)$ ließ sich andererseits bei der Einwirkung von $AlCl_3$ auf ein Gemisch von $Cr(C_6H_6)_2$ und $Cr(C_6H_5C_6H_5)_2$ glatt nachweisen [89].

Präparative Bedeutung haben Ligandenübertragungsreaktionen vor allem für die Synthese von Cyclobutadien-Metall-Komplexen erlangt. In diesem Fall kommt eine direkte Darstellung aus einer Metallverbindung und dem freien Ringliganden wegen der Instabilität des Cyclobutadiens und seiner Derivate nicht in Betracht. In Schema I (siehe nachfolgende Seite) sind die wesentlichen Ergebnisse der Untersuchungen von P. M. Maitlis und Mitarbeitern zusammengefaßt [140].

Die Isolierung der bei den Reaktionen der Tetraarylcyclobutadien-Palladium-Komplexe $[C_4R_4PdX_2]_2$ mit Metallcarbonylen ebenfalls zu erwartenden Palladiumcarbonyl-Verbindungen ist bisher nicht gelungen; an ihrer Stelle wird stets metallisches Palladium erhalten. Bei den Umsetzungen der Tetramethylcyclobutadien-Nickel-Komplexe mit $Co_2(CO)_8$ gemäß (55) und (56) konnte dagegen gezeigt werden, daß – unter optimalen Bedingungen – aus einem Molekül $[C_4Me_4NiX_2]_2$ zwei Moleküle $Ni(CO)_4$ entstehen [141].

$$[C_4Me_4NiCl_2]_2 + 3\,Co_2(CO)_8 \xrightarrow{\text{THF}} 2\,C_4Me_4Co_2(CO)_6 + 2\,Ni(CO)_4 + 2\,CoCl_2 + 4\,CO \qquad (55)$$

$$2\,[C_4Me_4NiJ_2]_2 + 3\,Co_2(CO)_8 \xrightarrow{\text{THF}} 4\,C_4Me_4Co(CO)_2J + 4\,Ni(CO)_4 + 4\,CoJ_2 \qquad (56)$$

Die analoge Übertragung eines C_5H_5-Liganden von $C_5H_5Fe(CO)_2Br$ (oder $[C_5H_5Fe(CO)_2]_2$) auf das Zentralatom eines C_4R_4-Ni- oder C_4R_4-Pd-Komplexes ist – wie in Schema I gezeigt – ebenfalls möglich. Dabei bilden sich die kationischen Vierring-Fünfring-Sandwich-Verbindungen $[C_4R_4MC_5H_5]^+$ [140]. In den Gl. (57)–(60) sind weitere Beispiele für C_5H_5-Ligandentransfer-Reaktionen angegeben [140].

$$1.5\text{-}C_8H_{12}PdBr_2 + C_5H_5Fe(CO)_2Br \longrightarrow [1.5\text{-}C_8H_{12}PdC_5H_5][FeBr_4] \qquad (57)$$

$$C_4Me_4Co(CO)_2J + [C_5H_5Fe(CO)_2]_2 \longrightarrow C_4Me_4CoC_5H_5 \qquad (58)$$

$$TiCl_4 + [C_5H_5Fe(CO)_2]_2 \longrightarrow C_5H_5TiCl_3 \qquad (59)$$

$$Fe(CO)_5 + (C_5H_5)_2TiCl_2 \longrightarrow Fe(C_5H_5)_2 \qquad (60)$$

Besonders interessant ist der Befund, daß auch *zwei* Ringliganden, und zwar gemäß (61), auf ein Metall übertragen werden können [142]. Die geringe Ausbeute von $C_4Ph_4CoC_5H_5$ deutet allerdings darauf hin, daß der Transfer nicht gleichzeitig sondern schrittweise erfolgt.

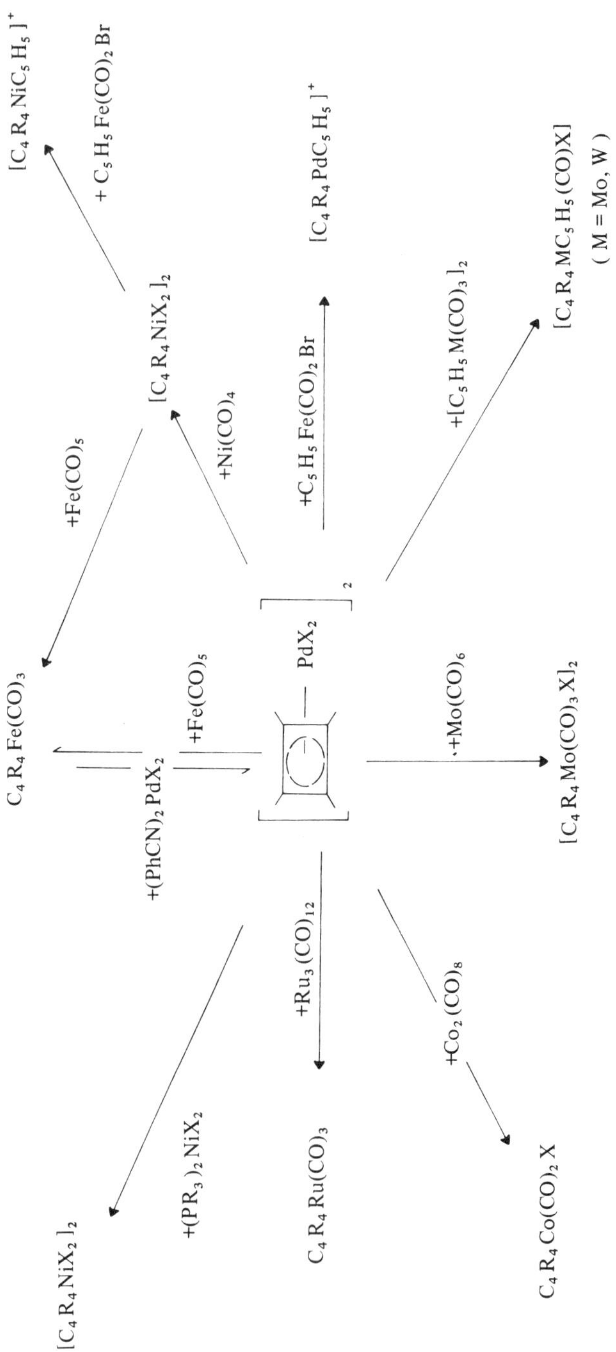
[C4 R4 NiC5 H5]+
+ C5 H5 Fe(CO)2 Br
[C4 R4 NiX2]2
+Ni(CO)4
+Fe(CO)5
[C4 R4 PdC5 H5]+
+C5 H5 Fe(CO)2 Br
+[C5 H5 M(CO)3]2
[C4 R4 MC5 H5 (CO)X]
(M = Mo, W)
PdX2
C4 R4 Fe(CO)3
+Fe(CO)5
+(PhCN)2 PdX2
+Mo(CO)6
[C4 R4 Mo(CO)3 X]2
+Ru3 (CO)12
+Co2 (CO)8
+(PR3)2 NiX2
[C4 R4 NiX2]2
C4 R4 Ru(CO)3
C4 R4 Co(CO)2 X

Schema 1

$$[C_4Ph_4PdC_5H_5]Br + Co_2(CO)_8 \longrightarrow C_4Ph_4CoC_5H_5 \qquad (61)$$

Bei den Reaktionen sowohl der C_5H_5- als auch der C_4R_4-Metall-Komplexe ist die intermediäre Bildung von freien C_nH_n-Liganden wenig wahrscheinlich. Maitlis [140] nimmt an, daß z.B. bei der Umsetzung von $[C_4R_4PdX_2]_2$ und $M(CO)_n$ im Primärschritt ein 1 : 1-Addukt der Ausgangsverbindungen entsteht, was zu einer Lockerung der C_4R_4-Pd-Bindung führt. Unter Abspaltung von CO und Knüpfung der C_4R_4-M-Bindung könnte sich dann – möglicherweise über einen nach XX zu formulierenden Übergangszustand – das Endprodukt bilden.

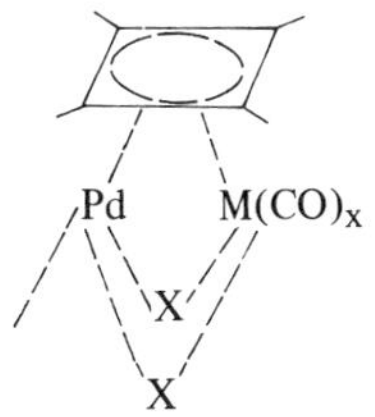

XX

Für den erfolgreichen Ablauf der hier diskutierten Ligandenverdrängungsreaktionen sind wahrscheinlich mehrere Einflüsse bestimmend. Die thermodynamische Stabilität der Ausgangs- und Endverbindungen (die zugleich die Stärke der Ring-Metall-Bindungen reflektiert) spielt dabei sicher eine wichtige Rolle. Auf die Bedeutung kinetischer Faktoren weist folgende Tatsache hin: Metallhexacarbonyle $M(CO)_6$ (M = Cr, Mo, W) reagieren mit $[C_4R_4PdX_2]_2$ wesentlich langsamer als $Ni(CO)_4$ oder $Co_2(CO)_8$; die Ausbeute an dem entstehenden Cyclobutadien-Komplex $[C_4R_4M(CO)_3X]_2$ ist außerdem sehr gering [140]. Da für die Spaltung einer M-CO-Bindung in $M(CO)_6$ eine sehr viel höhere Energie als für die Spaltung einer Ni-CO- oder Co-CO-Bindung in dem entsprechenden Metallcarbonyl notwendig ist, könnte es sein, daß auch die Energie des für die Umsetzung von $M(CO)_6$ und $[C_4R_4PdX_2]_2$ maßgebenden Übergangszustandes sehr viel höher liegt als diejenige des Übergangszustandes der Reaktion von $[C_4R_4PdX_2]_2$ und $Ni(CO)_4$ oder $Co_2(CO)_8$. Eine hohe Energiebarriere bedingt aber möglicherweise nicht nur einen Bruch der C_4R_4-Pd- sondern auch der C_4R_4-M-Bindung, d.h. es entsteht freies Cyclobutadien C_4R_4, das unter Dimerisierung zu einem Cyclooctatetraen C_8R_8 oder mit dem frei werdenden CO zu einem tetrasubstituierten Cyclopentadienon C_5R_4O reagiert. Beide Verbindungen werden bei vielen Ligandenübertragungsreaktionen der Cyclobutadien-Komplexe in wechselnden Mengen gefaßt. Schließlich sei noch darauf hingewiesen, daß auch der bereitwillige Ablauf der Reaktion von $[C_4R_4PdX_2]_2$ mit $Ni(PR'_3)_2X_2$ (einer Gruppe quadratisch-planarer Ni(II)-Komplexe) auf kinetische Einflüsse (wenig gehinderte Wechselwirkung des Nickels mit einem C_4R_4-Liganden) zurückgeführt werden könnte.

5. Schlußbemerkungen

Das Bild, das sich bei einer zusammenfassenden Betrachtung der Ringliganden-verdrängungsreaktionen von Aromaten-Metall-Komplexen im jetzigen Zeitpunkt bietet, mutet recht heterogen, in verschiedener Hinsicht vielleicht auch sehr fragmentarisch an. Der vorliegende Überblick will auf keinen Fall diesen Eindruck verwischen. Er ist vielmehr als ein Zwischenbericht zu sehen, der weiteres Interesse wecken und zu neuen Untersuchungen anreizen soll. Von der Vielzahl der Fragen, die sich bei einer Diskussion der bis jetzt vorliegenden Ergebnisse aufdrängen, seien hier nur zwei noch herausgestellt:

Warum reagiert Dicyclopentadienylnickel mit Lewis-Basen wesentlich bereitwilliger als das strukturanaloge Ferrocen?

Was ist der Grund dafür, daß bei der Umsetzung eines Aromaten-Metall-Halbsandwichkomplexes, wie z.B. $C_6H_6Cr(CO)_3$, mit Lewis-Basen sehr viel leichter eine Spaltung der Metall-Ring-Bindung als bei der entsprechenden Reaktion eines Sandwichkomplexes, wie z.B. $Cr(C_6H_6)_2$, erfolgt?

Bei der Beantwortung der ersten Frage könnte man zunächst versucht sein, den Reaktivitätsunterschied mit bindungstheoretischen Argumenten zu erklären. $Ni(C_5H_5)_2$ ist im Gegensatz zu $Fe(C_5H_5)_2$ paramagnetisch und hat zwei ungepaarte Elektronen. Sowohl MO-Berechnungen [143–148] als auch die genaue Analyse des Elektronenspektrums [145,149,150] führen zu der Aussage, daß $Ni(C_5H_5)_2$ einen $^3A_{2g}$-Grundzustand besitzt, wobei die energetische Reihenfolge der obersten besetzten MO's $e^*_{1g} > a_{1g} > e_{2g}$ ist (siehe auch Schema 2). Die beiden ungepaarten Elektronen befinden sich danach in einem Molekülorbital der Symmetrie e^*_{1g}, das antibindenden Charakter zeigt. Als Folge davon sollte – im Vergleich zu dem diamagnetischen Ferrocen, in dem das e^*_{1g}-Orbital nicht besetzt ist – im Fall des Dicylopentadienylnickels eine Schwächung der Metall-Ring-Bindung resultieren, die auch durch verschiedene, unabhängig durchgeführte Messungen bestätigt wird. So ergibt sich z.B. aus IR-Untersuchungen [151] ein sehr beträchtlicher Unterschied in der Kraftkonstante k der Metall-Ring-Bindung [$Fe(C_5H_5)_2$: k = 2,7 mdyn/Å; $Ni(C_5H_5)_2$: k = 1,5 mdyn/Å], was mit dem deutlich geringeren M-Ring-Abstand im $Fe(C_5H_5)_2$ (1,65 Å) [152] gegenüber $Ni(C_5H_5)_2$ (1,79 Å) [143,153] in Einklang steht. Die aus thermochemischen [154] und massenspektrometrischen Daten [155] ermittelten Bildungsenthalpien der Dicyclopentadienyl-Metallverbindungen stimmen ebenfalls mit der Annahme einer stabileren Metall-Ring-Bindung im Fall des Ferrocens überein.

Die Aussage, daß die Stärke der $M\text{-}C_5H_5$-Bindung beim Übergang von $Fe(C_5H_5)_2$ zu $Ni(C_5H_5)_2$ abnimmt, muß jedoch in erster Linie als ein thermodynamischer Aspekt gesehen werden, der *nicht unbedingt* mit einem Unterschied der Reaktivität in Beziehung stehen muß. Für das paramagnetische $Cr(C_5H_5)_2$ (das ebenso wie $Ni(C_5H_5)_2$ zwei ungepaarte Elektronen besitzt) wurde z.B. für die Kraftkonstante k der Metall-Ring-Bindung ein fast gleich

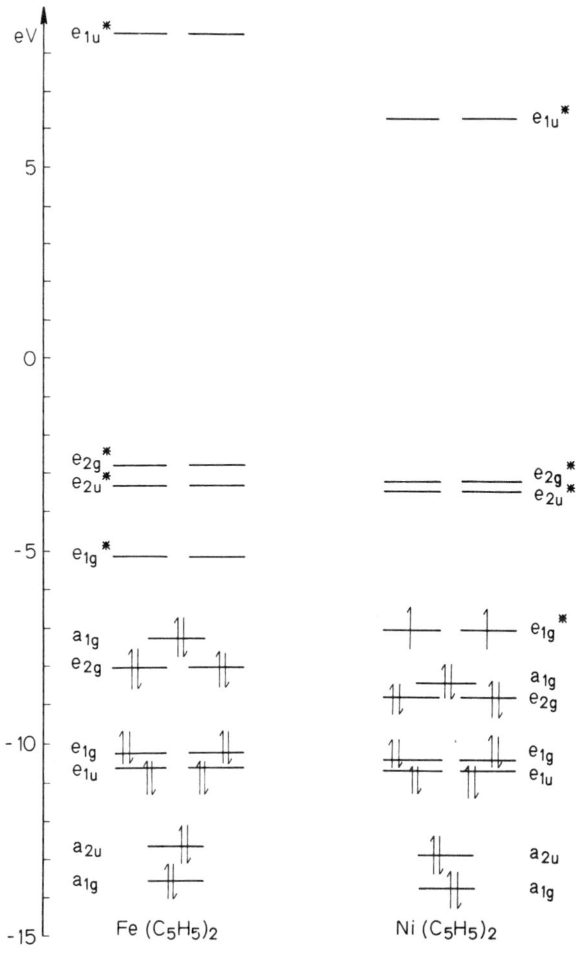

Schema 2 (Angaben nach Ref. 145)

großer Wert (1,6 mdyn/Å) wie für das Dicyclopentadienylnickel gefunden [151], obwohl – wie gerade eigene Untersuchungen in letzter Zeit gezeigt haben – $Cr(C_5H_5)_2$ *wesentlich* reaktionsträger als $Ni(C_5H_5)_2$ ist. Für die Reaktivität dieser Sandwich-Komplexe müssen also noch andere Faktoren als die vorwiegend thermodynamischen verantwortlich sein. Dabei ist es allerdings schwierig zu sagen, welcher Art diese Faktoren (die die Reaktionsgeschwindigkeit, d.h. die Größe von $\Delta G^{\neq}$ bestimmen) sind.

Auf die Bedeutung sterischer Einflüsse wurde bereits unter 3. aufmerksam gemacht. Da exakte Rechnungen oder auch nur zuverlässige Abschätzungen über die Höhe des Aktivierungsberges bei komplexen Reaktionen der hier diskutierten Art sehr schwierig (wenn nicht im Moment sogar noch unmöglich) sind, sei im Hinblick auf eine rein qualitative Deutung folgende Überlegung angestellt. Wenn man davon ausgeht, daß der geschwindigkeitsbestimmende Schritt bei den Ringligandenverdrängungsreaktionen der Sandwich- und Halbsandwich-Komplexe ein *bimolekularer* Schritt ist, so könnte für dessen Geschwindigkeit bei einem vorwiegend *nucleophilen* Angriff des substituierenden Liganden L die Energie der tiefsten unbesetzten MO's des Komplexes (die für eine Überlappung mit dem Donororbital von L vor allem in Betracht kommen) maßgebend sein. Bei einem vorwiegend *elektrophilen* Angriff von L wäre dann ganz analog die Energie der höchsten besetzten MO's des Komplexes (die dann mit dem Akzeptororbital von L überlappen) von Bedeutung. Eine Reaktion von $Fe(C_5H_5)_2$ oder $Ni(C_5H_5)_2$ mit einem bevorzugt nucleophilen Liganden L würde danach – und zwar in jedem Fall – im Übergangszustand zu einer Wechselwirkung des Donororbitals von L mit einem *antibindenden* MO der Sandwich-Verbindung führen. Diese Wechselwirkung sollte für das Dicyclopentadienylnickel wegen der geringeren Stabilität der $Ni\text{-}C_5H_5$Bindung schon im Grundzustand (siehe auch die in Schema 2 angegebene Elektronenbesetzung) leichter eine Bindungsspaltung bewirken. Bei der Umsetzung von $Fe(C_5H_5)_2$ oder $Ni(C_5H_5)_2$ mit einem bevorzugt elektrophilen Liganden L sollte im Fall des Nickel-Komplexes (einer „Elektronenüberschuß-Verbindung") eher eine Wechselwirkung des obersten MO's mit dem freien Akzeptororbital von L möglich sein als im Fall des Eisen-Komplexes. Dies hätte zur Folge, daß die Energie des Übergangszustandes bei der Umsetzung von $Ni(C_5H_5)_2$ niedriger wäre als bei der entsprechenden Reaktion von $Fe(C_5H_5)_2$. Dicyclopentadienylnickel sollte also mit L bereitwilliger als Ferrocen reagieren, was mit der Erfahrung übereinstimmt.

Könnte man auf der Grundlage dieser Überlegung auch den Reaktivitätsunterschied zwischen dem Halbsandwich-Komplex $C_6H_6Cr(CO)_3$ und dem Sandwich-Komplex $Cr(C_6H_6)_2$ verstehen? Eine eindeutige Antwort auf diese Frage ist im Moment wohl nicht möglich. Nach dem Ergebnis qualitativer und semiempirischer MO-Rechnungen (für $C_6H_6Cr(CO)_3$ siehe [156,157]; für $Cr(C_6H_6)_2$ siehe [147,158,159]) darf als gesichert gelten, daß in beiden – *diamagnetischen* – Verbindungen ausschließlich *bindende* Molekülorbitale besetzt sind. Zuverlässige Angaben über die Energie dieser MO's (und auch über die Energie der tiefsten unbesetzten MO's) liegen jedoch nicht vor. Ebenso fehlen Daten über die thermodynamische Stabilität der Metall-Ring-Bindung in Benzol-chrom-tricarbonyl, die mit den Werten der Kraftkonstante [151] und der Bildungsenthalpie [160,161] der $Cr\text{-}C_6H_6$-Bindungen in Dibenzolchrom verglichen werden könnten. Ein Hinweis auf eine größere thermodynamische Stabilität der Chrom-Benzol-Bindung in $Cr(C_6H_6)_2$ ist möglicherweise aus dem geringeren Abstandswert Cr-C = 2,13 Å [162] gegenüber Cr-C(Ring) = 2,23 Å in $C_6H_6Cr(CO)_3$ [163] zu entnehmen, obwohl eine solche Korrelation (vergleiche z.B. die Bildungsenthalpien [160,161] und Abstandswerte [162,164] von $Cr(C_6H_6)_2$ und $V(C_6H_6)_2$) auch wieder sehr problematisch erscheint.

Es bleibt zu hoffen, daß zukünftige Berechnungen eine klarere Aussage über den Zusammenhang zwischen Elektronenstruktur und Reaktivität eines Metallkomplexes geben können und dann auch zufriedenstellendere Antworten auf die hier gestellten Fragen möglich sind.

Dank

Für zahlreiche gewinnbringende Diskussionen möchte ich den Herren V. Harder und M. Textor, Anorganisch-chemisches Institut der Universität Zürich, bestens danken. Herrn Prof. Dr.-Ing. H. Behrens, Anorganisch-chemisches Institut der Universität Erlangen-Nürnberg, und Herrn Priv. Doz. Dr. M. Herberhold, Laboratorium für Anorganische Chemie der Technischen Universität München, bin ich für die Erlaubnis, unveröffentlichte Ergebnisse aus ihren Arbeitskreisen zu erwähnen, ebenfalls sehr verbunden.

Die in dem vorliegenden Überblick zitierten eigenen Untersuchungen wurden in großzügiger Weise von dem Schweizerischen Nationalfonds unterstützt.

6. Literatur

1) Fischer, E. O., Werner, H.: Metal π-Complexes, Vol. 1. Complexes with Di- and Oligoolefinic Ligands. Amsterdam: Elsevier 1966. – Bennett, M. A.: Advan. Organometal. Chem. *4,* 353 (1966). – Quinn, H. W., Tsai, J. H.: Advan. Inorg. Chem. Radiochem. *12,* 217 (1969).

2) Beckert, O.: Dissertation TH München 1957.

3) Pauson, P. L., Stubbs, W. H.: Angew. Chem. *74,* 466 (1962).

4) Behrens, H., Meyer, K.: Z. Naturforschg. *21b,* 489 (1966).

5) Olechowski, J. R., McAlister, C. G., Clark, R. F.: Inorg. Chem. *4,* 246 (1965).

6) Werner, H., Harder, V., Deckelmann, E.: Helv. Chim. Acta *52,* 1081 (1969).

7) Nixon, J. F.: Chem. Comm. *1966,* 34.

8) – J. Chem. Soc. (A) *1967,* 1136.

9) – Sexton, M. D.: J. Chem. Soc. (A) *1969,* 1089.

10) Jefferson, R., Klein, H. F., Nixon, J. F.: Chem. Comm. *1969,* 536.

11) Fischer, E. O., Werner, H.: Chem. Ber. *95,* 703 (1962).

12) Harder, V., Werner, H.: Unveröffentlichte Ergebnisse; – siehe Harder, V.: Dissertation Univ. Zürich 1972.

13) Gubin, S. P., Rubezhov, A. Z., Winch, B. L., Nesmeyanov, A. N.: Tetrahedron Letters *1964,* 2881.

14) Ustynyuk, Yu. A., Voevodskaya, T. I., Zharikova, N. A., Ustynyuk, N. A.: Dokl. Akad. Nauk SSSR *181,* 372 (1968); – Dokl. Chem. *181,* 640 (1968).

15) – Barinov, I. V., Voevodskaya, T. I., Rodionova, N. A.: Proceed. XIII. Internat. Conf. Coord. Chem., Cracow-Zakopane *1970,* 153.

16) Akker, M. van den, Jellinek, F.: Rec. Trav. Chim. *86,* 897 (1967); – Akker, M.: van den: Thesis Rijks-Universität Groningen 1970.

17) Salzer, A., Werner, H.: Unveröffentlichte Ergebnisse.

18) Sato, M., Sato, F., Yoshida, T.: J. Organometal. Chem. *27,* 273 (1971).

19) Schropp, W. K.: J. Inorg. Nucl. Chem. *24,* 1688 (1962).
20) Dobbie, R. C., Green, M., Stone, F. G. A.: J. Chem. Soc. (A) *1969,* 1881.
21) McClellan, W. R., Hoehn, H. H., Cripps, H. N., Muetterties, E. L., Howk, B. W.: J. Am. Chem. Soc. *83,* 1601 (1961).
22) McBride, D. W., Dudek, E., Stone, F. G. A.: J. Chem. Soc. *1964,* 1752.
23) Fischer, E. O., Beckert, O., Hafner, W., Stahl, H. O.: Z. Naturforschg. *10b,* 598 (1955).
24) Piper, T. S., Cotton, F. A., Wilkinson, G.: J. Inorg. Nucl. Chem. *1,* 165 (1955).
25) Fischer, E. O., Palm, C.: Chem. Ber. *91,* 1725 (1958).
26) Dubeck, M.: J. Am. Chem. Soc. *82,* 502 (1960).
27) Harbourne, D. A., Stone, F. G. A.: J. Chem. Soc. (A) *1968,* 1765.
28) Bruce, M. I., Iqbal, M. Z.: J. Organometal. Chem. *17,* 469 (1969).
29) Tilney-Bassett, J. F.: J. Chem. Soc. *1961,* 577.
30) Dubeck, M.: J. Am. Chem. Soc. *82,* 6193 (1960).
31) Dahl, L. F., Wei, C. H.: Inorg. Chem. *2,* 713 (1963).
32) McBride, D. W., Pruett, P. L., Pitcher, E., Stone, F. G. A.: J. Am. Chem. Soc. *84,* 497 (1962).
33) Kleiman, J. P., Dubeck, M.: J. Am. Chem. Soc. *85,* 1544 (1963).
34) Werner, H., Mattmann, G., Salzer, A., Winkler, T.: J. Organometal. Chem. *25,* 461 (1970).
35) Otsuka, S., Nakamura, A., Yoshida, T.: Inorg. Chem. *7,* 261 (1968).
36) Nesmeyanov, A. N., Vol'kenau, N. A., Bolesova, I. N.: Dokl. Akad. Nauk SSSR *149,* 615 (1963); Tetrahedron Letters *1963,* 1725.
37) – – Shivotzeva, L. S.: Dokl. Akad. Nauk SSSR *160,* 1327 (1965).
38) Nesmeyanov, A. N.: Pure Appl. Chem. *17,* 211 (1968).
39) Khand, I. U., Pauson, P. L., Watts, W. E.: J. Chem. Soc. (C) *1968,* 2257, 2261; *1969,* 116.
40) Nesmeyanov, A. N., Vol'kenau, N. A., Bolesova, I. N.: Dokl. Akad. Nauk SSSR *166,* 607 (1966).
41) Khand, I. U., Pauson, P. L., Watts, W. E.: J. Chem. Soc. (C) *1969,* 2024.
42) Bublitz, D. E.: Can. J. Chem. *42,* 2381 (1964); J. Organometal. Chem. *16,* 149 (1969).
43) Suvorova, O. N., Domrachev, G. A., Razuvaev, G. A.: Dokl. Akad. Nauk SSSR *183,* 850 (1968).
44) Pavlycheva, A. V., Domrachev, G. A., Razuvaev, G. A., Suvorova, O. N.: Dokl. Akad. Nauk SSSR *184,* 105 (1969).
45) Fischer, E. O., Hafner, W.: Z. Naturforschg. *10b,* 140 (1955).
46) – – Stahl, H. O.: Z. Anorg. Allg. Chem. *282,* 47 (1955).
47) – Jira, R.: Z. Naturforschg. *10b,* 355 (1955).
48) – – Z. Naturforschg. *9b,* 618 (1954).
49) – Plesske, K.: Chem. Ber. *91,* 2719 (1958).
50) Reynolds, L. T., Wilkinson, G.: J. Inorg. Nucl. Chem. *9,* 86 (1958).
51) Britisches Patent 781065 (1952); US-Patent 2818417 (1955), Ethyl Corporation.
52) Pruett, R. L., Morehouse, E. L.: Chem. Ind. *1958,* 980.
53) Fischer, E. O., Wirzmüller, A.: Z. Naturforschg. *12b,* 737 (1957).
54) Green, M. L. H., Wilkinson, G.: J. Chem. Soc. *1958,* 4314.
55) Fischer, E. O., Hafner, W.: Z. Naturforschg. *9b,* 503 (1954).
56) – Vigoureux, S.: Chem. Ber. *91,* 2205 (1958).
57) – Löchner, A.: Z. Naturforschg. *15b,* 266 (1960).
58) Murray, J. G.: J. Am. Chem. Soc. *81,* 752 (1959).
59) Piper, T. S., Wilkinson, G.: J. Inorg. Nucl. Chem. *2,* 38 (1956).
60) King, R. B., Bisnette, M. B.: Inorg. Chem. *3,* 791 (1964).

61) Kruck, Th., Hieber, W., Lang, W.: Angew. Chem. *78*, 208 (1966); Angew. Chem., Internat. Ed. *5*, 247 (1966).
62) Harder, V., Müller, J., Werner, H.: Helv. Chim. Acta *54*, 1 (1971).
63) Joh, T., Hagihara, N., Murahashi, S.: Bull. Chem. Soc. Japan *40*, 661 (1967).
64) Behrens, H., Brandl, H., Lutz, K.: Z. Naturforschg. *22b*, 99 (1967).
65) – – Z. Naturforschg. *22b*, 1353 (1967).
66) – Aquila, W.: Z. Naturforschg. *22b*, 454 (1967).
67) – – Unveröffentlichte Ergebnisse; – siehe Aquila, W.: Dissertation Univ. Erlangen-Nürnberg 1967.
68) Sellmann, D.: Z. Naturforschg. *25b*, 1482 (1970).
69) Behrens, H., Schindler, H.: Z. Naturforschg. *23b*, 1110 (1968).
70) Brunner, H., Wachsmann, H.: J. Organometal. Chem. *15*, 409 (1968).
71) Mawby, R. J., White, C.: Chem. Comm. *1968*, 312; Inorg. Chim. Acta *4*, 261 (1970).
72) Hart-Davis, A. J., White, C., Mawby, R. J.: Inorg. Chim. Acta *4*, 441 (1970).
73) King, R. B., Stokes, J. C., Korenowski, T. F.: J. Organometal. Chem. *11*, 641 (1968).
74) Podall, H., Giraitis, A. P.: J. Org. Chem. *26*, 2587 (1961).
75) Herberhold, M.: Persönliche Mitteilung.
76) James, T. A., McCleverty, J. A.: J. Chem. Soc. (A) *1970*, 3318.
77) Locke, J., McCleverty, J. A.: Inorg. Chem. *5*, 1157 (1966).
78) James, T. A., McCleverty, J. A.: J. Chem. Soc. (A) *1970*, 3308.
79) King, R. B., Bisnette, M. B.: Inorg. Chem. *6*, 469 (1967).
80) Howe, J. J., Pinnavaia, T. J.: J. Am. Chem. Soc. *92*, 7342 (1970).
81) Fischer, E. O., Öfele, K.: Chem. Ber. *90*, 2532 (1957).
82) Kruck, Th.: Z. Naturforschg. *19b*, 165 (1964); Chem. Ber. *97*, 2018 (1964).
83) – Prasch, A.: Z. Naturforschg. *19b*, 669 (1964).
84) Chatt, J., Watson, H. R.: Proc. Chem. Soc. *1960*, 243.
85) Behrens, H., Meyer, K., Müller, A.: Z. Naturforschg. *20b*, 74 (1965).
86) Calderazzo, F., Cini, R.: J. Chem. Soc. *1965*, 818.
87) Fischer, E. O.: Angew. Chem. *69*, 715 (1957).
88) – Seeholzer, J.: Z. Anorg. Allg. Chem. *312*, 244 (1961).
89) Hein, F., Kartte, K.: Z. Anorg. Allg. Chem. *307*, 22, 52, 89 (1960).
90) Fischer, E. O., Breitschaft, S.: Angew. Chem. *75*, 94 (1963); Angew. Chem., Internat. Ed. *2*, 44 (1963); Chem. Ber. *99*, 2905 (1966).
91) Fischer, E. O., Breitschaft, S.: Chem. Ber. *96*, 2451 (1963).
92) Poilblanc, R., Bigorgne, M.: Bull. Soc. Chim. France *1962*, 1301.
93) Zingales, F., Chiesa, A., Basolo, F.: J. Am. Chem. Soc. *88*, 2707 (1966).
94) Nicholls, B., Whiting, M. C.: J. Chem. Soc. *1959*, 551.
95) Pidcock, A., Smith, J. D., Taylor, B. W.: J. Chem. Soc. (A) *1967*, 872.
96) – – – J. Chem. Soc. (A) *1969*, 1604.
97) Matthews, C. N., Magee, T. A., Wotiz, J. H.: J. Am. Chem. Soc. *81*, 2273 (1959); Magee, T. A., Matthews, C. N., Wang, T. S., Wotiz, J. H.: J. Am. Chem. Soc. *83*, 3200 (1961).
98) Jones, C. E., Coskran, K. J.: Inorg. Chem. *10*, 55 (1971).
99) Fowles, G. W. A., Jenkins, D. K.: Inorg. Chem. *3*, 257 (1964).
100) Abel, E. W., Bennet, M. A., Wilkinson, G.: J. Chem. Soc. *1959*, 2323.
101) Dieck, H. tom, Renk, I.W., Brehm, H. P.: Z. Anorg. Allg. Chem. *379*, 169 (1970).
102) Brown, D. A., Cunningham, D., Glass, W. K.: Chem. Comm. *1966*, 306.
103) Connelly, N. G., Dahl, L. F.: Chem. Comm. *1970*, 880.
104) Natta, G., Ercoli, R., Calderazzo, F., Santambrogio, F.: Chim. Ind. (Milano) *40*, 1003 (1958); Ercoli, R., Calderazzo, F., Alberola, A.: Chim. Ind. (Milan) *41*, 975 (1959).
105) Mangini, A., Taddei, F.: Inorg. Chim. Acta *2*, 8, 12 (1968).

106) Jackson, W. R., Nicholls, B., Whiting, M. C.: J. Chem. Soc. *1960,* 469.
107) Strohmeier, W., Mittnacht, H.: Chem. Ber. *93,* 2085 (1960).
108) – Hobe, D.v.: Z. Naturforschg. *18b,* 981 (1963).
109) Nyholm, R. S., Snow, M. R., Stiddard, M. H. B.: J. Chem. Soc. *1965,* 6564.
110) Stiddard, M. H. B., Townsend, R. E.: J. Chem. Soc. (A) *1969,* 2355.
111) Anker, M. W., Cotton, R., Tomkins, I. B.: Pure Appl. Chem. *18,* 23 (1968).
112) Fischer, E. O., Louis, E., Kreiter, C. G.: Angew. Chem. *81,* 397 (1969); Angew. Chem., Internat. Ed. *8,* 377 (1969).
113) Werner, H., Prinz, R., Deckelmann, E.: Chem. Ber. *102,* 95 (1969).
114) Deckelmann, E., Werner, H.: Helv. Chim. Acta *52,* 892 (1969).
115) Werner, H., Deckelmann, E.: Chimia *23,* 195 (1969).
116) – Heckl, B., Harder, V.: Chimia *25,* 242 (1971).
117) Strohmeier, W., Mittnacht, H.: Z. Physik. Chem., N.F. *29,* 339 (1961).
118) – Staricco, E.H.: Z. Physik. Chem., N.F. *38,* 315 (1963).
119) – Müller, R.: Z. Physik. Chem., N.F. *40,* 85 (1964).
120) Gracey, D. E. F., Henbest, H. B., Jackson, W. R., McMullen, C. H.: Chem. Comm. *1965,* 566.
121) Pidcock, A., Taylor, B. W.: J. Chem. Soc. (A) *1967,* 877.
122) – Smith, J. D., Taylor, B. W.: Inorg. Nucl. Chem. Letters *4,* 467 (1968); Inorg. Chem. *9,* 638 (1970).
123) Werner, H., Deckelmann, E., Deckelmann, K.: Vortragsberichte zum Symposium „Koordinationschemie der Übergangselemente", Jena 1969, Sektion C, S. 202.
124) Für Übersichtsartikel hierzu siehe: Brown, D. A.: Inorg. Chim. Acta Reviews *1,* 35 (1967); Angelici, R. J.: Organometal. Chem. Reviews *3,* 173 (1968); Strohmeier, W., in: Fortschritte der chemischen Forschung, Band 10/2, S. 306. Berlin-Heidelberg-New York: Springer 1968. – Werner, H.: Angew. Chem. *80,* 1017 (1968); Angew. Chem., Internat. Ed. *7,* 930 (1968).
125) Fischer, R. D.: Chem. Ber. *93,* 165 (1960).
126) Werner, H., Deckelmann, E., Deckelmann, K.: Unveröffentlichte Ergebnisse; – siehe Deckelmann, K.: Dissertation Univ. Zürich 1971.
127) Randall, E. W., Sutton, L. E.: Proc. Chem. Soc. *1959,* 469.
128) Fischer, E. O., Schreiner, S.: Chem. Ber. *92,* 938 (1959).
129) Strohmeier, W., Hellmann, H.: Ber. Bunsenges. Physik. Chem. *67,* 190 (1963).
130) Goan, J. C., Berg, E., Podall, H. E.: J. Org. Chem. *29,* 975 (1964); Rosenblum, M., Fish, R. W., Bennett, C.: J. Am. Chem. Soc. *86,* 5166 (1964); Hetnarski, B.: Bull. Acad. Polon. Sci. *13,* 515, 523, 557, 563 (1965); Watanabe, H., Motoyama, I., Hata, K.: Bull. Chem. Soc. Japan *39,* 850 (1966); Adman, E., Rosenblum, M., Sullivan, S., Margulis, T. N.: J. Am. Chem. Soc. *89,* 4540 (1967).
131) Fitch, J. W., Lagowski, J. J.: Inorg. Chem. *4,* 864 (1965).
132) Huttner, G., Fischer, E. O., Fischer, R. D., Carter, O. L., McPhail, A. T., Sim, G. A.: J. Organometal. Chem. *6,* 288 (1966); Carter, O. L., McPhail, A. T., Sim, G. A.: J. Chem. Soc. (A) *1966,* 822; Fitch, J. W., Lagowski, J. J.: J. Organometal. Chem. *5,* 480 (1966); Huttner, G., Fischer, E. O.: J. Organometal. Chem. *8,* 299 (1967).
133) Langford, C. H., Gray, H. B.: Ligand Substitution Processes. New York – Amsterdam: W. A. Benjamin Inc. 1965.
134) Tilney-Bassett, J. F.: Proc. Chem. Soc. *1960,* 419; J. Chem. Soc. *1963,* 4784.
135) Schroll, G. E.: US-Patent 3054815 (1962); Chem. Abstr. *58,* 1494c (1963).
136) Yamazaki, H., Nishido, T., Matsumoto, Y., Sumida, S., Hagihara, N.: J. Organometal. Chem. *6,* 86 (1966).
137) Rausch, M. D., Chang, Y. F., Gordon, H. B.: Inorg. Chem. *8,* 1355 (1969).
138) Gorsich, R. D.: J. Am. Chem. Soc. *82,* 4211 (1960).

139) Morrison, I. G., Pauson, P. L.: Proc. Chem. Soc. *1962*, 177.

140) Maitlis, P. M.: Adv. Organometal. Chem. *4*, 95 (1966); Annals of the New York Academy of Sciences *159*, 110 (1969); Pollock, D. F., Maitlis, P. M.: J. Organometal. Chem. *26*, 407 (1971).

141) Bruce, R., Maitlis, P. M.: Can. J. Chem. *45*, 2011, 2017 (1967).

142) Maitlis, P. M., Efraty, A., Games, M. L.: J. Organometal. Chem. *2*, 284 (1964); J. Am. Chem. Soc. *87*, 719 (1965).

143) Dunitz, J. D., Orgel, L. E.: J. Chem. Phys. *23*, 954 (1955).

144) Shustorovich, E. M., Dyatkina, M. E.: Zh. Neorgan. Khim. *3*, 2721 (1958); *4*, 553 (1959); *6*, 1247 (1961).

145) Schachtschneider, J. H., Prins, R., Ros, P.: Inorg. Chim. Acta *1*, 462 (1967).

146) Armstrong, A. T., Carroll, D. G., McGlynn, S. P.: J. Chem. Phys. *47*, 1104 (1967).

147) Prins, R., Voorst, J. D. W. van: J. Chem. Phys. *49*, 4665 (1968).

148) Rettig, M. F., Drago, R. S.: J. Am. Chem. Soc. *91*, 1361 (1969); *91*, 3432 (1969).

149) Scott, D. R., Becker, R. S.: J. Chem. Phys. *35*, 516 (1961); J. Organometal. Chem. *4*, 409 (1965).

150) Pavlik, I., Cerny, V., Maxova, E.: Collection Czech. Chem. Commun. *35*, 3045 (1970).

151) Fritz, H. P.: Chem. Ber. *92*, 780 (1959); Adv. Organometal. Chem. *1*, 239 (1964).

152) Siebold, E. A., Sutton, L. E.: J. Chem. Phys. *23*, 1967 (1955); Dunitz, J. D., Orgel, L. E., Rich, A.: Acta Cryst. *9*, 373 (1956); Bohn, R. K., Haaland, A.: J. Organometal. Chem. *5*, 470 (1966).

153) Weiss, E., Fischer, E. O.: Z. Anorg. Allg. Chem. *278*, 219 (1955).

154) Wilkinson, G., Pauson, P. L., Cotton, F. A.: J. Am. Chem. Soc. *76*, 1970 (1954).

155) Friedman, L., Irsa, A. P., Wilkinson, G.: J. Am. Chem. Soc. *77*, 3689 (1955); Müller, J., D'Or, L.: J. Organometal. Chem. *10*, 313 (1967).

156) Carroll, D. G., McGlynn, S. P.: Inorg. Chem. *7*, 1285 (1968).

157) Brown, D. A., Rawlinson, R. M.: J. Chem. Soc. (A) *1969*, 1534.

158) Shustorovich, E. M., Dyatkina, M. E.: Dokl. Akad. Nauk SSSR *128*, 1234 (1959); Zh. Strukt. Khim. *2*, 40 (1961).

159) Fischer, R. D.: Theoret. Chim. Acta *1*, 418 (1963).

160) Reckziegel, A.: Dissertation Univ. München 1962.

161) Müller, J., Göser, P.: J. Organometal. Chem. *12*, 163 (1968); Herberich, G. E., Müller, J.: J. Organometal. Chem. *16*, 111 (1969).

162) Jellinek, F.: Nature *187*, 871 (1960) – Cotton, F. A., Dollase, W. A., Wood, J. S.: J. Am. Chem. Soc. *85*, 1543 (1963) – Weiss, E., Fischer, E. O.: Z. Anorg. Allg. Chem. *286*, 142 (1956). – Ibers, J. A.: J. Chem. Phys. *40*, 3129 (1964). – Keulen, E., Jellinek, F.: J. Organometal. Chem. *5*, 490 (1966).

163) Corradini, P., Allegra, G.: J. Am. Chem. Soc. *81*, 2271 (1959). – Bailey, M. F., Dahl, L. F.: Inorg. Chem. *4*, 1314 (1965).

164) Fischer, E. O., Fritz, H. P., Manchot, J., Priebe, E., Schneider, R.: Chem. Ber. *96*, 1418 (1963).

165) Anmerkung bei der Korrektur: Kinetische Daten der Reaktion von $Ni(C_5H_5)_2$ mit RSH (R = C_6H_5, p-ClC_6H_4, p-$CH_3C_6H_4$, C_6H_{11}, $C_6H_5CH_2$, $C_{10}H_7$) sind kürzlich publiziert worden; Ellgen, P. C., Gregory, C. D.: Inorg. Chem. *10*, 980 (1971).

166) Anmerkung bei der Korrektur: Über Umsetzungen von $[C_5H_5NiCO]_2$ mit Lewis-Basen, die unter Disproportionierung verlaufen, berichten ebenfalls: Ellgen, P. C.: Inorg. Chem. *10*, 232 (1971). – Carty, A. J., Efraty, A., Ng, T. W.: Can. J. Chem. *47*, 1429 (1969). – King, R. B.: Inorg. Chem. *2*, 936 (1963).

Eingegangen am 14. Juni 1971